AS UNIT 1

STUDENT GUIDE

CCEA

Chemistry

Basic concepts in physical and inorganic chemistry

Alyn G. McFarland

HODDER
EDUCATION
AN HACHETTE UK COMPANY

Content Guidance

■ Formulae, equations and amounts of substance

Writing chemical formulae

Writing chemical formulae is an important skill and mistakes made in equations are often due to incorrect formulae. Many covalent compounds you come across have to be learned or can be worked out from the name or, in organic chemistry, the formulae are determined from a general formula.

Formulae of ionic compounds from the ions

The formula of an ionic compound can be determined simply from the charges on the ions, since the overall charge on an ionic compound must be zero. For example:

- metal ions include sodium ion (Na^+), magnesium ion (Mg^{2+}), copper(II) ion (Cu^{2+}), iron(III) ion (Fe^{3+})
- simple non-metal ions include chloride ion (Cl^-), oxide ion (O^{2-}), nitride ion (N^{3-})

Metal ions are positively charged and the positive charge is the same as the group number for groups I, II and III metals: for example, Na^+, Mg^{2+}, Ca^{2+}, Al^{3+}.

Metals in other groups, such as lead and tin which are in group IV, form 2+ ions — e.g. lead(II) chloride and tin(II) oxide. These metals can form other compounds such as lead(IV) chloride and tin(IV) bromide, but these are covalent compounds.

Simple non-metal ions are negatively charged and the charge is 8 − group number.

You can work out the charge on a transition metal ion from the name of the compound, e.g. in copper(II) oxide the copper(II) ion is Cu^{2+}; in silver(I) chloride, the silver(I) ion is Ag^+; in iron(III) chloride the iron(III) ions are Fe^{3+}. Most transition metals can exist as 2+ ions.

Some ionic compounds are made up of **molecular ions**. Common examples are shown in Table 1.

Table 1 Examples of molecular ions

Ion	Formula	Ion	Formula
Sulfate	$SO_4{}^{2-}$	Phosphate	$PO_4{}^{3-}$
Sulfite	$SO_3{}^{2-}$	Chlorate	$ClO_3{}^-$
Thiosulfate	$S_2O_3{}^{2-}$	Hypochlorite	OCl^-
Hydrogensulfate	$HSO_4{}^-$	Hydroxide	OH^-
Hydrogencarbonate	$HCO_3{}^-$	Dichromate	$Cr_2O_7{}^{2-}$
Carbonate	$CO_3{}^{2-}$	Chromate	$CrO_4{}^{2-}$
Nitrate	$NO_3{}^-$	Permanganate	$MnO_4{}^-$
Nitrite	$NO_2{}^-$	Ammonium	$NH_4{}^+$

Exam tip
Do not give the charge on an ion as, for example, Mg^{+2} — this would not be accepted. Charge should be just + or −, or a number followed by the charge, such as 2+, 3−. Later in this unit you will meet oxidation states, which should be stated as + or − followed by a number.

A **molecular ion** is two or more atoms covalently bonded with an overall charge.

Exam tip
Learn the formulae (including charges) of common molecular ions. These will not be given in the data leaflet. You may be asked to determine the formula of a molecular ion including its charge from the formula of a compound containing the ion.

Worked examples

1 Sodium chloride contains Na^+ and Cl^- ions.
   One of each ion is required so sodium chloride is NaCl.

Remember that chemical formulae are written without the charges.

2 Calcium chloride contains Ca^{2+} and Cl^- ions.
   Two Cl^- ions are required for one Ca^{2+} ion so calcium chloride is $CaCl_2$.

3 Magnesium oxide contains Mg^{2+} and O^{2-} ions.
   One of each ion is required so magnesium oxide is MgO.

4 Copper(II) hydroxide contains Cu^{2+} and OH^- ions.
   Two OH^- ions are required for one Cu^{2+} ion so copper(II) hydroxide is $Cu(OH)_2$.

5 Ammonium sulfate contains NH_4^+ and SO_4^{2-} ions.
   Two NH_4^+ ions are required for one SO_4^{2-} ion so ammonium sulfate is $(NH_4)_2SO_4$.

6 Aluminium nitrate contains Al^{3+} and NO_3^- ions.
   Three NO_3^- ions for one Al^{3+} ion so aluminium nitrate is $Al(NO_3)_3$.

Exam tip

The hydroxide ion is a unit and so when we have more than one of them we must use brackets. This is true of all molecular ions.

Exam tip

Mistakes commonly made with formulae include hydroxides of metal ions with valency greater than 1, carbonates and sulfates of metal ions with valency of 1, ammonium carbonate and sulfate.

- Brackets are often left out, e.g. calcium hydroxide is often *incorrectly written* as $CaOH_2$ instead of $Ca(OH)_2$.
- The two metal or ammonium ions may not be included in the formula, e.g. potassium sulfate is often *incorrectly written* as KSO_4 instead of K_2SO_4.

Examples of *correct* sulfate and carbonate formulae are: Na_2CO_3, Na_2SO_4, K_2CO_3, $(NH_4)_2SO_4$.

Knowledge check 1

Write the chemical formula for ammonium dichromate.

Formula of an ion from a compound

This is a compound type of question where the formula of an ion including its charge is determined from the formula of an ionic compound. You may be asked to determine the formulae of other compounds containing the ion.

Worked example

The formula of bismuth(III) ethanedioate has the formula $Bi_2(C_2O_4)_3$. What is the formula of the ethanedioate ion?

Answer

The bismuth(III) ion is Bi^{3+} and there are two of them. There are three C_2O_4 ions and since the charges have to balance, the charge on the C_2O_4 ion is 2−. The ethanedioate ion is $C_2O_4^{2-}$.

Equations for chemical reactions

A balanced symbol equation shows the rearrangement of atoms in a chemical reaction. You will be asked to write equations for reactions that are familiar to you but you may also be asked for some that are unfamiliar.

Worked example

Ammonium carbonate decomposes on heating to produce ammonia, carbon dioxide and water. Write an equation for this reaction.

Answer

Ammonium carbonate is the compound of the ammonium ion (NH_4^+) and the carbonate ion (CO_3^{2-}). So ammonium carbonate is $(NH_4)_2CO_3$.

Unbalanced equation:

$$(NH_4)_2CO_3 \rightarrow NH_3 + CO_2 + H_2O$$

Balanced equation:

$$(NH_4)_2CO_3 \rightarrow 2NH_3 + CO_2 + H_2O$$

$2NH_3$ are required to balance the atoms on either side. Remember, you can never change a formula to balance an equation.

Ionic equations

Ionic equations will be encountered in oxidation, reduction, redox and precipitation reactions later in this guide.

An ionic equation is an equation where spectator ions (ions that do not take part in the reaction) are removed and only the ions that take part in the reaction are included. The total charge in any ionic equation should be equal on both sides.

Worked example 1

For the reaction:

$$AgNO_3(aq) + NaCl(aq) \rightarrow AgCl(s) + NaNO_3(aq)$$

a white precipitate of silver(I) chloride is observed — shown as a solid in the equation above. The Na^+ and NO_3^- ions remain in solution and do not take part in the formation of the white silver(I) chloride. The ionic equation is:

$$Ag^+(aq) + Cl^-(aq) \rightarrow AgCl(s)$$

State symbols are often important in ionic equations since the product can be a precipitate (a solid).

Exam tip

You will meet balanced equations throughout the course and this is a fundamental skill in chemistry. Make sure that you can balance equations.

Knowledge check 2

Write an equation for the reaction of magnesium nitride with water to form magnesium hydroxide and ammonia gas.

Worked example 2

For the reaction:

$$Na_2CO_3(aq) + 2HCl(aq) \rightarrow 2NaCl(aq) + H_2O(l) + CO_2(g)$$

the sodium ions (Na^+) and the chloride ions (Cl^-) are spectator ions as they remain in solution. The ionic equation for the reaction is:

$$CO_3^{2-} + 2H^+ \rightarrow H_2O + CO_2$$

Relative atomic mass, relative molecular mass and molar mass

The **relative formula mass** (RFM) is calculated by adding the **relative atomic masses** of all the atoms in a formula.

1 mole of a substance is the RFM measured in grams.

1 mole of Mg = 24 g	(RAM of Mg = 24)
1 mole of N_2 = 28 g	(RAM of N = 14)
1 mole of CO_2 = 44 g	(RAM of C = 12, O = 16)
1 mole of $Cu(OH)_2$ = 98 g	(RAM of H = 1, O = 16, Cu = 64)
1 mole of $Fe(NO_3)_3$ = 242 g	(RAM of N = 14, O = 16, Fe = 56)

The term **molar mass** is also used. The units of molar mass are g/mol or $g\,mol^{-1}$.

RFM (relative formula mass) is often used to represent RAM and RMM.

Molar mass of $Cu(OH)_2$ = $98\,g\,mol^{-1}$ Molar mass of N_2 = $28\,g\,mol^{-1}$

The mole and the Avogadro constant

A balanced symbol equation for a reaction gives the rearrangement of the atoms in a chemical reaction.

The equation:

$$Cu + S \rightarrow CuS$$

can be read as: one Cu atom reacts with one S atom to form one CuS unit.

However, the masses of atoms are too small to measure, so the number of particles used in measurements is scaled up by 6.02×10^{23}. This number is called the **Avogadro constant**, which is represented by the symbols N_A or L.

One Cu atom has a mass of 1.063×10^{-22} g, so 6.02×10^{23} Cu atoms have a mass of approximately 64 g.

Avogadro's constant is quoted as $6.02 \times 10^{23}\,mol^{-1}$, since this is the number of particles in 1 mole.

The term 'amount' is the quantity, which is measured in **moles** (or mol as a unit).

Relative atomic mass (RAM) is the average (weighted mean) mass of an atom of an element relative to one-twelfth of the mass of an atom of carbon-12.

The **relative molecular mass** (RMM) is the average (weighted mean) mass of a molecule relative to one-twelfth of the mass of an atom of carbon-12.

The **relative formula mass** (RFM) is the average (weighted mean) of a species relative to one-twelfth of the mass of an atom of carbon-12.

Molar mass is the mass of 1 mole of a substance.

The **Avogadro constant** is defined as the number of atoms in 12.000 g of carbon-12.

A **mole** of a substance is the amount of a substance that contains the Avogadro constant (6.02×10^{23}) number of atoms, molecules or groups of ions.

This equation:

$$Cu + S \rightarrow CuS$$

can be read as: 1 mole of copper atoms reacts with 1 mole of sulfur atoms to form 1 mole of copper(II) sulfide

Calculating amount, in moles

The amount of a substance is measured in moles, often represented by n. It is calculated from the mass using the expression:

$$\text{amount, in moles, } n = \frac{\text{mass (g)}}{\text{RFM}}$$

This expression can be rearranged to calculate mass in grams from amount, in moles (n) and RFM:

$$\text{mass (g)} = n \times \text{RFM}$$

Or RFM may be calculated from mass and amount, in moles (n):

$$\text{RFM} = \frac{\text{mass (g)}}{n}$$

Worked example

Calculate the amount, in moles, present in 2.71 g of carbon dioxide. Give your answer to three significant figures.

Answer

$$\text{amount, in moles} = \frac{\text{mass (g)}}{\text{RFM}} = \frac{2.71}{44} = 0.0616 \, \text{mol}$$

Calculating amount, in moles, in a solution

The concentration of any solution is measured in mol dm^{-3}. A solution of concentration $1 \, \text{mol dm}^{-3}$ will have 1 mol of solute dissolved in $1 \, \text{dm}^3$.

The amount, in moles, of a solute in a solution may be calculated using the expression:

$$\text{amount, in moles } (n) = \frac{\text{volume (cm}^3) \times \text{concentration (mol dm}^{-3})}{1000}$$

Worked example

Calculate the amount, in moles, of sodium hydroxide present in $15.0 \, \text{cm}^3$ of a solution of concentration $1.25 \, \text{mol dm}^{-3}$. Give your answer to three significant figures.

Answer

$$\text{amount, in moles } (n) = \frac{\text{volume (cm}^3) \times \text{concentration (dm}^{-3})}{1000}$$
$$= \frac{15.0 \times 1.25}{1000} = 0.0188 \, \text{mol}$$

Exam tip

The mol unit is often forgotten. Always include units where they are needed with a numerical answer.

Questions may require an answer to a specific number of decimal places or significant figures. Make sure you know the difference, and give the answer to the correct level asked for. If this is not detailed, it is usually best to use three significant figures for numbers less than one, or two decimal places for answers greater than one.

Exam tip

Units are very important in chemistry. $1 \, \text{dm}^3$ is the same as 1 litre but chemists prefer $1 \, \text{dm}^3$. $1 \, \text{cm}^3$ is the same as 1 millilitre. There are $1000 \, \text{cm}^3$ in $1 \, \text{dm}^3$.

Exam tip

For both recent examples you are asked to give the answer to three significant figures. This is common when all the initial data are given to three significant figures. Always give a numerical answer to the number of significant figures required by the question.

Using the Avogadro constant

Calculations involving number of particles or mass of a certain number of particles involve using the Avogadro constant. N_A or L represents the Avogadro constant, which is equal to $6.02 \times 10^{23}\,\text{mol}^{-1}$.

$$\frac{\text{mass (g)}}{\text{RFM}} = \text{amount, in moles} = \frac{\text{number of particles}}{N_A}$$

These can be rearranged to give:

$$\text{mass (g)} = \text{moles} \times \text{RFM} \qquad \text{and} \qquad \text{number of particles} = \text{moles} \times N_A$$

Worked example 1

Calculate the mass of 100 zinc atoms (the Avogadro constant $= 6.02 \times 10^{23}\,\text{mol}^{-1}$).

Answer

$$\text{amount (in moles) for 100 zinc atoms} = \frac{100}{6.02 \times 10^{23}} = 1.66 \times 10^{-22}\,\text{mol}$$

$$\text{mass of 100 zinc atoms} = 1.66 \times 10^{-22} \times 65 = 1.08 \times 10^{-20}\,\text{g}$$

Worked example 2

Calculate the number of hydrogen atoms present in 0.125 g of methane, CH_4.

Answer

$$n = \frac{\text{mass (g)}}{\text{RFM}} = \frac{0.125}{16} = 7.813 \times 10^{-3}\,\text{mol} \times N_A\,(6.02 \times 10^{23}\,\text{mol}^{-1})$$

$$= 4.70 \times 10^{21}\,\text{molecules of CH}_4$$

Each CH_4 contains four H atoms, so number of H atoms $= 4 \times 4.70 \times 10^{21} = 1.88 \times 10^{22}$ atoms of H.

Interpreting balanced symbol equations quantitatively

The balanced symbol equation is the key to calculations. Follow these steps:

- Using the mass of one of the reactants, which should be given to you, calculate the amount, in moles, of this substance.
- Using the balancing numbers in the equation, calculate the amount, in moles, of whatever substance you are asked to calculate.
- Change the amount, in moles, of this substance to mass (or volume if required).

Several expressions are required to help you do this:

Expression 1: $\text{amount (in moles)} = \dfrac{\text{mass (g)}}{\text{RFM}}$

Expression 2: $\text{mass (g)} = \text{amount (in moles)} \times \text{RFM}$

Knowledge check 3

Calculate the amount, in moles, of potassium carbonate present in 2.08 g. Give your answer to three significant figures.

Knowledge check 4

Calculate the number of atoms in 1.00 g of argon gas. Give your answer to three significant figures.

Exam tip

Masses may be given in various units, e.g. mg, kg or tonnes. To convert between mg and g, multiply by 10^{-3}; to convert between kg and g multiply by 10^3; to convert between tonnes and g multiply by 10^6.

Mass is often written more simply as m. Amount, in moles, is often written as n. Hence the above equations can be learnt in a simplified form as long as you remember what the abbreviations are:

Expression 1: $n = \dfrac{\text{mass}}{\text{RFM}}$

Expression 2: $m = n \times \text{RFM}$

Worked example

5.21 g of calcium carbonate are heated to constant mass. What mass of calcium oxide would be obtained? Give your answer to three significant figures.

Answer

$$CaCO_3(s) \rightarrow CaO(s) + CO_2(g)$$

5.21 g of calcium carbonate is the information given. So using expression 1 this can be converted to moles:

$CaCO_3$ RFM = 100

$$n = \frac{\text{mass}}{\text{RFM}} \Rightarrow n = \frac{5.21}{100} = 0.0521 \text{ moles of } CaCO_3$$

In the balanced symbol equation given, there are no balancing numbers so this means that 1 mol of $CaCO_3$ gives 1 mol of CaO and 1 mol of CO_2. So 0.0521 mol of $CaCO_3$ gives 0.0521 mol of CaO and 0.0521 mol of CO_2.

To calculate the mass of CaO formed in this reaction, we must use expression 2:

CaO RFM = 56

$$\text{mass} = n \times \text{RFM} = 0.0521 \times 56 = 2.92 \text{ g}$$

2.92 g of calcium oxide are formed.

Limiting reactant and excess reactant

Limiting reactant is the reactant in a chemical reaction that limits the amount of product that can be formed. The reaction will stop when all of the limiting reactant is consumed.

Excess reactant is the reactant in a chemical reaction that remains when a reaction stops because the limiting reactant has been completely consumed. The excess reactant remains because there is nothing with which it can react.

Figure 1 shows what happens when complete motorbikes are assembled with an excess of wheels.

Exam tip

The heating to constant mass in this question refers to that fact that we want all of the 5.21 g of calcium carbonate to decompose. With the release of carbon dioxide into the atmosphere, the solid will decrease in mass. If we repeatedly heat and check the mass, we will know that all of the calcium carbonate has been decomposed when the mass does not change in successive measurements. This is called heating to constant mass.

Knowledge check 5

Calculate the mass of silver formed when 1.42 g of silver(I) nitrate are heated to constant mass.

$$2AgNO_3 \rightarrow 2Ag + O_2 + 2NO_2$$

Give your answer to 3 significant figures.

Step 3: Determine which is the limiting reactant and which is the excess reactant.

3333.33 mol of C are present, so CaO is the limiting reactant (as there is not enough of it to react with all 3333.33 mole of C) and C is in excess (as there is more of it than can react with all of the CaO).

Step 4: Using the limiting reactant for calculation. The limiting reactant is now used because it is the one that determines the other number of moles of reactants used and the number of moles of products formed in the reaction.

$$CaO + 3C \rightarrow CaC_2 + CO$$

mol 714.29   2142.87   714.29   714.29

714.29 mol of CaO forms 714.29 mol of CaC_2

Mass of calcium carbide formed = 714.29×64 = 45 714.56 g = 45.715 kg

(RFM of CaC_2 = 64)

Water of crystallisation

Many salts (formed from acids) when they are solid are **hydrated**.

If hydrated salts are heated to constant mass in an open container (so the water vapour can escape) all of the water of crystallisation is removed, leaving an **anhydrous salt**.

Hydrated salts are written with the water of crystallisation — for example, $CuSO_4.5H_2O$ or $CoCl_2.6H_2O$ or $Na_2CO_3.10H_2O$.

The number of moles of water of crystallisation attached to 1 mol of the salt is called the degree of hydration. Many hydrated salts effloresce when left in the open. This means they lose their water of crystallisation gradually to the atmosphere. Heating in an open container removes the water of crystallisation more rapidly. Heating a hydrated salt to constant mass will remove all of the water of crystallisation (Figure 2).

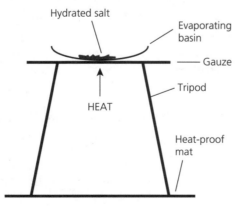

Figure 2 Removing water of crystallisation

The degree of hydration can be determined by taking mass measurements before heating and after heating to constant mass.

A **hydrated salt** is one that contains **water of crystallisation**.

Water of crystallisation is water chemically bonded within a crystal structure.

An **anhydrous salt** contains no water of crystallisation.

Exam tip

Figure 2 shows how the hydrated salt may be heated in an evaporating basin. The process of heating to constant mass may be asked about in terms of the steps you would take — you would heat and weigh and repeat this process until the mass no longer changes.

Questions often ask what initial weighings you would make in this type of procedure. You should weigh the empty container and the mass of the container with the hydrated salt.

Method of calculation

In any question where you are asked to determine the degree of hydration from mass data or percentage data, follow the flow chart in Figure 3.

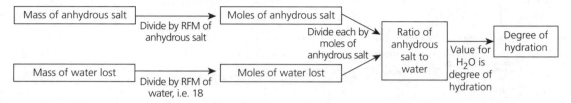

Figure 3 Calculating the degree of hydration

Use the data given to determine the mass of the anhydrous salt and the mass of water lost. From these masses, divide by the RFM and determine the moles of each. The hydrated salt breaks up as shown below:

hydrated salt $\rightarrow$ anhydrous salt + nH$_2$O

1 mole   1 mole n moles

The ratio of the anhydrous salt to the water is 1:n. So convert the moles to a simple ratio in which the anhydrous salt is 1 and the value for water is the degree of hydration.

Worked example 1

A sample of 5.72 g of hydrated sodium carbonate, Na$_2$CO$_3$.xH$_2$O, contains 2.12 g of anhydrous sodium carbonate. From the given data, the masses and moles are determined as follows:

Mass of anhydrous salt = 2.12 g

Mass of water = 5.72 − 2.12 = 3.6 g

From the data given, the masses and the moles are determined (Figure 4).

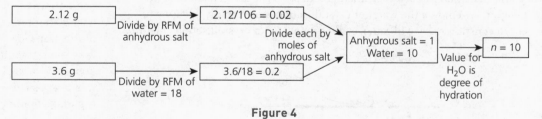

Figure 4

Alternative method

In this type of calculation the RFM of the hydrated salt may be determined instead. The number of moles of the hydrated salt are the same as the number of moles of the anhydrous salt.

So 0.02 mol of Na$_2$CO$_3$ formed from 0.02 mol of Na$_2$CO$_3$.xH$_2$O.

Using the mass of the hydrated salt (5.72 g) and the number of moles (0.02 mol), the RFM is calculated as 5.72/0.02 = 286.

Subtract the mass of the Na$_2$CO$_3$ (106), leaving 180.

This 180 is due to the mass of the water, so divide by the RFM of water (18) = 180/18 = 10. So x = 10.

Worked example 2

Hydrated nickel(II) sulfate, $NiSO_4.xH_2O$, contains 41.1% water. Determine the value of x.

Answer

The salt contains 58.9% (100 − 41.1) anhydrous nickel(II) sulfate.

Mass of water = 41.1 g

Mass of anhydrous nickel(II) sulfate = 58.9 g

From the data given, the masses and the moles are determined (Figure 5).

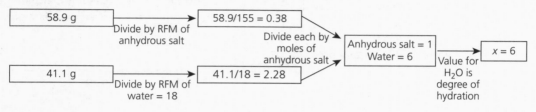

Figure 5

Worked example 3

An older sample of hydrated sodium carbonate, $Na_2CO_3.xH_2O$, was heated to constant mass in an evaporating basin. The following mass measurements were made.

Mass of evaporating basin = 53.07 g (1)

Mass of evaporating basin and hydrated sodium carbonate = 58.52 g (2)

Mass of evaporating basin and solid after heating to constant mass = 55.43 g (3)

So mass of anhydrous salt = (3) − (1) = 55.43 − 53.07 = 2.36 g

And mass of water lost = (2) − (3) = 58.52 − 55.43 = 3.09 g

From the data given, the masses and moles are determined as shown in Figure 6.

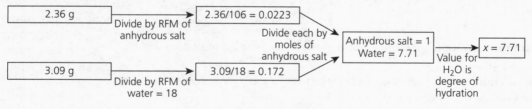

Figure 6

Exam tip

The degree of hydration may not be a whole number. Hydrated salts lose their water of crystallisation over time. The degree of hydration of older salts may be an average across all of the sample. This is common when mass measurements are given in a practical situation.

Knowledge check 7

6.16 g of hydrated magnesium sulfate, $MgSO_4.xH_2O$, are heated to constant mass. The residue has a mass of 3.01 g. Determine the value of x.

Summary

- For a solid:

  amount, in moles = $\dfrac{\text{mass (g)}}{\text{RFM}}$

- One mole of a substance is the mass in grams which contains Avogadro's constant number of particles (6.02×10^{23}).

- For a solution:

  amount, in moles =
  $$\dfrac{\text{volume (cm}^3) \times \text{concentration (mol dm}^{-3})}{1000}$$

- A balanced symbol equation is the ratio of the number of moles of each reactant and product.

- In a chemical reaction one or more reactants is often in excess and there will be one reactant, described as the limiting reactant, which limits the number of moles of product formed based on this number of moles of reactant.

- Heating a hydrated compound to constant mass produces an anhydrous compound.

- The simplest ratio of the number of moles of the anhydrous compound to the number of moles of water determines the degree of hydration.

Atomic structure

Basics of atomic structure

Atoms are composed of three subatomic particles: protons, neutrons and electrons.

In all of the following sections the term 'relative' is used. It is used as a comparison between particles as the masses and charges of these particles are so small that it is easier to use a standard measure and compare the rest to them (Table 2).

Table 2

Subatomic particle	Relative mass	Relative charge	Location in atom
Proton	1	+1	Nucleus
Neutron	1	0	Nucleus
Electron	$\dfrac{1}{1840}$	−1	Energy levels

- The **atomic number** of an element is the same as the number of protons in (the nucleus of) an atom.
- The **mass number** of a particular atom is the total number of protons and neutrons in (the nucleus of) an atom.
- **Relative atomic mass** (RAM) is the average (weighted mean) mass of an atom of an element relative to one-twelfth the mass of an atom of carbon-12.
- **Relative isotopic mass** (RIM) is the mass of an atom of an isotope of an element relative to one-twelfth the mass of an atom of carbon-12.
- **Relative molecular mass** (RMM) is the mass (weighted mean) of one molecule relative to one-twelfth the mass of an atom of carbon-12.
- **Relative formula mass** (RFM) is the average (weighted mean) mass of a formula unit relative to one-twelfth the mass of an atom of carbon-12.

The atomic number is often called the proton number. The relative molecular mass is calculated from the total of all the relative atomic masses in a single molecule.

Exam tip

It is the atomic number that defines the identity of the particle. A particle with 17 protons is always a chlorine particle — it may be a chlorine atom or a chloride ion depending on the number of electrons.

Isotopes are atoms that have the same atomic number but a different mass number (they contain the same number of protons but a different number of neutrons).

RMM is used for molecular covalent elements and compounds. RFM may be used for everything.

The number of subatomic particles in an atom or ion can be determined from the atomic number, mass number and the charge on the particle.

The atomic number is always the same as the number of protons in the nucleus.

The mass number is the sum of the number of protons and the number of neutrons; so subtracting the atomic number from the mass number gives the number of neutrons in an atom or an ion.

Atoms are electrically neutral as they have equal numbers of protons and electrons. Simple ions are charged particles formed when atoms lose or gain electrons. The number of electrons subtracted from the number of protons gives the charge. Remember an atom has no overall charge.

Number of protons = atomic number

Number of neutrons = mass number – atomic number

Charge = number of protons – number of electrons

Chlorine has two isotopes: chlorine-35 and chlorine-37. These are often written ^{35}Cl and ^{37}Cl. The relative atomic mass of chlorine is an average mass of the atoms taking the proportions in which they occur into account. 75% of all chlorine atoms are ^{35}Cl and 25% are ^{37}Cl.

The relative atomic mass of an element can be calculated from the relative isotopic masses of the isotopes (which are the same as the mass numbers) and the relative proportions in which they occur.

$$\text{RAM} = \frac{\Sigma(\text{mass of isotope} \times \text{relative abundance})}{\Sigma \text{relative abundance}}$$

where Σ represents the 'sum of' for all isotopes.

Worked example 1

Chlorine exists as two isotopes, ^{35}Cl and ^{37}Cl, which occur in the relative proportions 75% and 25% respectively. Calculate the relative atomic mass of chlorine.

Answer

$$\text{RAM} = \frac{(75 \times 35) + (25 \times 37)}{(75 + 25)} = \frac{3550}{100} = 35.5$$

Worked example 2

Determine the number of subatomic particles present in an aluminium ion, Al^{3+}.

Answer

From the periodic table, the atomic number of aluminium is 13 and the relative atomic mass is 27. Apart from chlorine (relative atomic mass 35.5), the relative atomic mass can be taken as the mass number of the most common isotope.

Exam tip

The symbol for an element may be written with its atomic number and mass number, i.e. $^{A}_{Z}E$ where E is the symbol for the element, Z is the atomic number and A is the mass number, e.g. $^{35}_{17}$Cl, $^{12}_{6}$Cl. The atomic number is often left out for isotopes, e.g. ^{35}Cl and ^{37}Cl.

Knowledge check 8

State the relative charge and mass of an electron, a proton and a neutron.

Number of protons = atomic number

so number of protons = 13

Number of neutrons = mass number − atomic number

so number of neutrons = 27 − 13 = 14

Charge = number of protons − number of electrons

+3 = 13 − number of electrons

so number of electrons = 13 − (+3) = 10

Mass spectrometry

A mass spectrometer is an analytical instrument used to determine the mass of atoms and molecules. A mass spectrometer atomises and ionises a sample, producing ions with a single positive charge. It is assumed that all the ions in a mass spectrum have a single positive charge. The data obtained from a mass spectrometer may be for an element or a compound.

If the data are for an element, the spectrum will show the masses and relative abundance for all the isotopes of the element. These data may be supplied in the form of a table or a mass spectrum, which has peaks that show the relative abundance of each of the isotopes.

Worked example

An element was analysed using a mass spectrometer. The spectrum showed that there were four isotopes. The relative isotopic masses and relative abundances are given in Table 3.

Table 3

Relative isotopic mass	Relative abundance
50	7
52	74
53	15
54	4

Calculate the relative atomic mass of the element to one decimal place.

Answer

This is carried out by thinking that in this sample 7 atoms have a mass of 50, 74 atoms have a mass of 52, 15 atoms have a mass of 53 and 4 atoms have a mass of 54. The relative atomic mass is simply the average mass of the atoms in the sample.

$$\text{RAM} = \frac{(7 \times 50) + (74 \times 52) + (15 \times 53) + (4 \times 54)}{7 + 74 + 15 + 4} = 52.1 \text{ to 1 d.p.}$$

Exam tip

Check the question for the number of decimal places and answer accordingly.

These data could have been supplied in a spectrum such as the one shown in Figure 7 with peaks at 50, 52, 53 and 54 of the heights shown as the relative abundance. The calculation is the same.

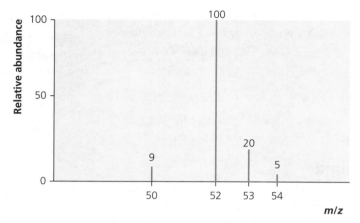

Figure 7 Mass spectrum of an element

The horizontal axis of a mass spectrum may be labelled 'mass' or 'mass to charge ratio' or 'm/e' or 'm/z' but it is still mass. The vertical axis can be relative abundance or percentage abundance.

You can be asked to identify the species that causes the peak at a particular mass value. For example, the peak at 54 in the mass spectrum is caused by $^{54}Cr^+$. Remember to include the mass number; all species are assumed to have a single positive charge.

Mass spectrum of a diatomic element

For a diatomic element such as chlorine there would be five peaks in the spectrum. Figure 8 shows the mass spectrum for chlorine. The species responsible for the peaks are shown in Table 4.

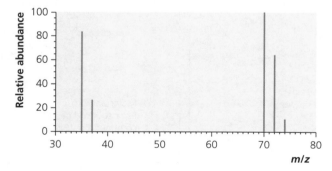

Figure 8 Mass spectrum of chlorine

Table 4

m/z	Species
35	$^{35}Cl^+$
37	$^{37}Cl^+$
70	$(^{35}Cl-^{35}Cl)^+$
72	$(^{35}Cl-^{37}Cl)^+$
74	$(^{37}Cl-^{37}Cl)^+$

^{35}Cl and ^{37}Cl occur naturally in a ratio of 3:1. This explains why the relative abundance of the peaks at m/z values of 35 and 37 are in the ratio 3:1. The peaks at m/z values of 70, 72 and 74 are in the ratio 9:6:1. This can be explained using probability. The probability of a chlorine atom being ^{35}Cl is $\frac{3}{4}$, whereas the probability of a chlorine atom being ^{37}Cl is $\frac{1}{4}$. The probability of obtaining two ^{35}Cl atoms in a molecule is $\frac{3}{4} \times \frac{3}{4} = \frac{9}{16}$. The probability of obtaining one ^{35}Cl and one ^{37}Cl is $2 \times (\frac{1}{4} \times \frac{3}{4}) = 6/16$. The fraction here is multiplied by 2 as it could be $^{35}Cl–^{37}Cl$ or $^{37}Cl–^{35}Cl$. The probability of obtaining two ^{37}Cl is $\frac{1}{4} \times \frac{1}{4} = \frac{1}{16}$. The top numbers in these fractions give the ratio of 9:6:1. For bromine, ^{79}Br and ^{81}Br occur in a 1:1 ratio and the relative abundance of the peaks at 158, 160 and 162 are in the ratio 1:2:1.

Mass spectrum of a compound

The mass spectrum for a compound is more complicated because the molecule breaks up during the process. The last major peak in the mass spectrum of a compound is called the molecular ion peak. The mass value for the molecular ion peak is the same as the RMM of the compound. The pattern seen at m/z values below the molecular ion peak is called the fragmentation pattern and each peak is caused by fragments of the molecule with a single positive charge. The fragmentation pattern for a compound is unique and may be used to identify a compound along with the RMM from the molecular ion peak.

Figure 9 shows the spectrum for the compound ethanol, CH_3CH_2OH. The species responsible for the peak at 46 is $CH_3CH_2OH^+$. The peak with the highest relative abundance in any mass spectrum is called the base peak.

Exam tip

Even though the relative atomic mass of bromine is 80, there is no isotope of bromine with a relative isotopic mass of 80.

Knowledge check 9

What is a molecular ion in mass spectrometry?

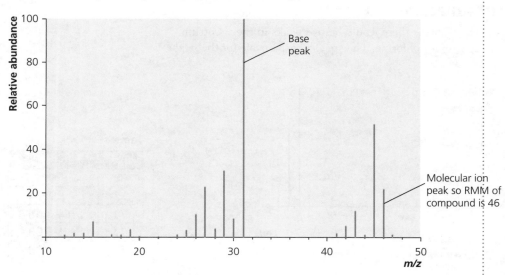

Figure 9 Mass spectrum of ethanol

Electronic configuration

Electrons are arranged in energy levels (the equivalent of shells) in which the energy of the electrons increases with increasing distance from the nucleus. The energy levels are labelled $n = 1$ (closest to the nucleus), $n = 2$, $n = 3$ etc. Energy levels are subdivided into subshells, which are made up of **orbitals**. Each orbital can be occupied by two electrons.

■ An 's subshell' is made up of one s orbital.

■ A 'p subshell' is made up of three p orbitals.

■ A 'd subshell' is made up of five d orbitals.

At $n = 1$ there is only an s subshell, at $n = 2$ there is an s subshell and a p subshell and at $n = 3$ there is an s subshell, a p subshell and a d subshell. There is a fourth subshell called an f subshell, but this is not required at this level.

An **orbital** is a region within an atom that can hold up to two electrons with opposite spin.

Table 5 The four types of orbitals — s, p, d and f

Type	Shape	Start at which energy level	Number of this type of orbital in a subshell	Maximum number of electrons
s	Spherical	1	1	2
p	Dumbbell	2	3	6
d	Not required	3	5	10
f	Not required	4	7	14

In each single orbital, electrons spin in opposite directions to minimise repulsions. Opposite spin is represented as ↑↓ in 'electron-in-box' diagrams.

The subshells fill in the following order: 1s, 2s, 2p, 3s, 3p, 4s, 3d, 4p

but the subshells should be written in the following order: 1s, 2s, 2p, 3s, 3p, 3d, 4s, 4p

Electrons are lost from subshells in the following order: 4p, 4s, 3d, 3p, 3s, 2p, 2s, 1s

The 2+ ion for transition metals is the most common and this is caused by loss of $4s^2$ electrons (not $3d$).

Electrons fill in order of energy levels and orbitals closest to the nucleus. The ground state is the term used to describe the electronic configuration in which all the electrons are in the lowest (available) energy levels.

Determining electronic configuration of atoms and ions

An iron atom

- Atomic number of iron = 26, so an iron atom has 26 protons.
- Atoms are electrically neutral, so an atom of iron has 26 electrons.
- Using order of filling: $1s^2, 2s^2, 2p^6, 3s^2, 3p^6, 4s^2, 3d^6$
- Remember that, when writing the order, the $4s$ has to come after the $3d$.
- So the electronic configuration of an iron atom is: $1s^2, 2s^2, 2p^6, 3s^2, 3p^6, 3d^6, 4s^2$.

An iron(II) ion

- The electronic configuration of an iron atom is: $1s^2, 2s^2, 2p^6, 3s^2, 3p^6, 3d^6, 4s^2$
- Remember that transition metal atoms lose their $4s$ electrons first.
- An iron atom loses two electrons to form a 2+ ion.
- So the electronic configuration of an iron(II) ion is: $1s^2, 2s^2, 2p^6, 3s^2, 3p^6, 3d^6$

A bromide ion, Br⁻

- The electronic configuration of a bromine atom is determined first.
- Atomic number of bromine = 35, so a Br atom has 35 protons and 35 electrons.
- The electronic configuration of a Br atom is: $1s^2, 2s^2, 2p^6, 3s^2, 3p^6, 3d^{10}, 4s^2, 4p^5$
- A bromine atom gains one electron to form a bromide ion.
- The electronic configuration of a bromide ion is: $1s^2, 2s^2, 2p^6, 3s^2, 3p^6, 3d^{10}, 4s^2, 4p^6$

The electronic configurations of chromium and copper atoms are unusual:

Cr: $1s^2, 2s^2, 2p^6, 3s^2, 3p^6, 3d^5, 4s^1$ *not* $3d^4, 4s^2$

Cu: $1s^2, 2s^2, 2p^6, 3s^2, 3p^6, 3d^{10}, 4s^1$ *not* $3d^9, 4s^2$

Note that Cr and Cu have unusual electronic configurations owing to the stability of the half-filled and filled d^5 and d^{10} configurations. However, the formation of ions of copper and chromium works in the normal way with the loss of the $4s$ electrons first.

There are three main blocks in the periodic table and these are based on the subshell in which the outer electrons are located (Figure 10).

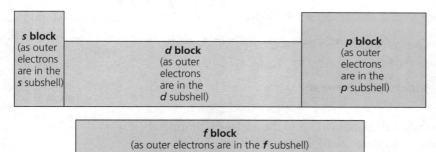

Figure 10 Blocks of the periodic table

Exam tip

As soon as you start any AS or A2 exam, put a star (*) at Cr and Cu on your periodic table to remind you that their electronic configurations are different from what you would expect. This can be asked in any AS or A2 unit.

Electron-in-box diagrams

The electronic configuration can be asked for in written format or in an 'electron-in-box' form. Electrons only pair when no other space is available in the subshell. Electrons in a subshell that are not paired spin in the same direction and are represented by arrows pointing in the same direction.

Figure 11

Often something like Figure 11 is used. Sometimes you have to label the subshells, so you will have to be able to identify and label $1s$, $2s$, $2p$ etc. It is relatively easy as the s subshell has only one orbital, the p subshell has three orbitals and the d subshell has five orbitals. Also, there is an s subshell at each energy level; there is a p subshell at each energy level except $n = 1$; there is a d subshell at each energy level except $n = 1$ and $n = 2$. Remember that the $4s$ is at a slightly lower energy level than the $3d$.

Worked examples

A nitrogen atom (atomic number 7)

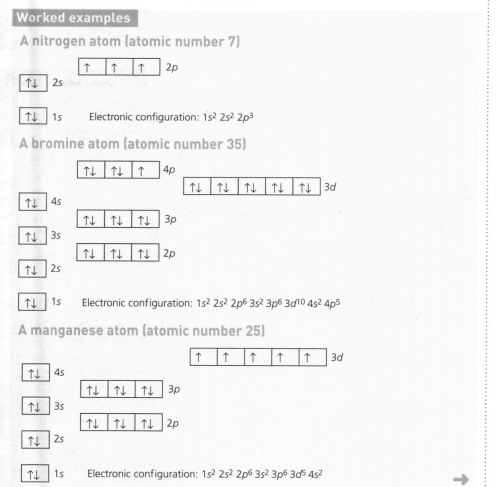

Exam tip

When filling the subshells remember to start at the lowest and work up. Also remember that for p and d subshells there should only be one electron in each subshell until each is half-filled and then start pairing electrons in subshells.

Finally remember that electrons are shown as arrows pointing up (↑) and arrows pointing down (↓). The opposite directions represent the different directions of spin of the electrons in one orbital. All the first set of electrons in one subshell should be in the same direction.

A copper atom (atomic number 29)

Remember that the electronic configuration of a copper atom includes $3d^{10}$ and not $3d^9$.

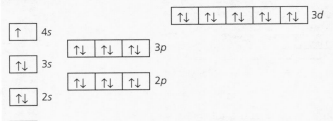

Electronic configuration: $1s^2\ 2s^2\ 2p^6\ 3s^2\ 3p^6\ 3d^{10}\ 4s^1$

An iron(III) ion

Electronic configuration of an iron atom is: $1s^2, 2s^2, 2p^6, 3s^2, 3p^6, 3d^6, 4s^2$

Electronic configuration of an iron(III) ion is: $1s^2, 2s^2, 2p^6, 3s^2, 3p^6, 3d^5$

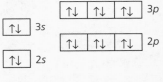

Electronic configuration: $1s^2\ 2s^2\ 2p^6\ 3s^2\ 3p^6\ 3d^5$

Ionisation energy

Values of ionisation energies are always endothermic and always measured in $kJ\,mol^{-1}$.

- **First ionisation energy** is the energy required to convert 1 mol of gaseous atoms into gaseous ions with a single positive charge.
- **Second ionisation energy** is the energy required to convert 1 mol of gaseous ions with a single positive charge into ions with a double positive charge.
- **Third ionisation energy** is the energy required to convert 1 mol of gaseous ions with a double positive charge into ions with a triple positive charge.

In the following example, X represents any element.

Example equation for first ionisation energy:

$$X(g) \rightarrow X^+(g) + e^-$$

Example equation for second ionisation energy:

$$X^+(g) \rightarrow X^{2+}(g) + e^-$$

Example equation for third ionisation energy:

$$X^{2+}(g) \rightarrow X^{3+}(g) + e^-$$

Exam tip

Remember that transition metal atoms lose their 4s electrons first.

Exam tip

You may be given all subshells in an 'electron-in-box' style question up to and including 4p, but you may only need to use some of them

Knowledge check 10

Explain why magnesium is described as being an s block element.

Exam tip

Equations for ionisation energy must have *gaseous* species (atoms or ions) and only 1 mol of electrons removed each time. The electrons do not require a state symbol.

$X(g) \rightarrow X^{2+}(g) + 2e^-$ would be a combination of the first and second ionisation energies. The value for this would be the first and second ionisation energies added together, for example:

First ionisation energy of magnesium = $+740\,\text{kJ}\,\text{mol}^{-1}$

This is for the change $Mg(g) \rightarrow Mg^+(g) + e^-$

Second ionisation energy of magnesium = $+1500\,\text{kJ}\,\text{mol}^{-1}$

This is for the change $Mg^+(g) \rightarrow Mg^{2+}(g) + e^-$

The energy for the change $Mg(g) \rightarrow Mg^{2+}(g) + 2e^- = +740 + 1500 = +2240\,\text{kJ}$

Patterns in first ionisation energy

Figure 12 shows the change in first ionisation energies from hydrogen (atomic number 1) to krypton (atomic number 36).

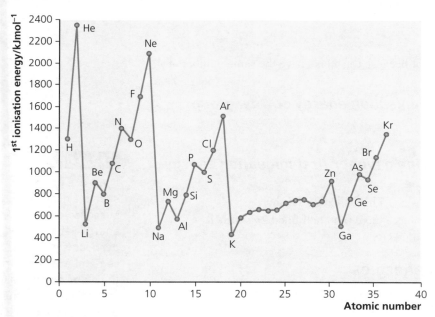

Figure 12 First ionisation energies of elements 1 to 36

Identifying elements

The elements that have the lowest first ionisation energy in each period are the alkali metals (group I).

The elements that have the highest first ionisation energy in each period are the noble gases (group VIII or 0).

Exam tip

You need to recognise the patterns in this graph and be able to explain them, as well as being able to identify elements in the graph.

Exam tip

Any element can be identified from the graph because the lowest first ionisation energies in any period are the group I elements. Count across the period from here. Note the general increase in first ionisation energy across any period and general decrease down any group.

Factors used to explain changes in ionisation energy

The following four factors are used to explain trends in ionisation energies:

1 nuclear charge
2 atomic radius
3 shielding (by inner electrons)
4 stability of filled and half-filled subshells

Trends seen in the graph of first ionisation energies

■ Across a period — first ionisation energy generally increases.
■ Down a group — first ionisation energy decreases.

Note that elements in groups II, V and 0 exhibit higher than expected first ionisation energy values owing to the stability of half-filled and filled subshells.

Reasons for increase in ionisation energy across a period

■ Increase in nuclear charge
■ Decrease in atomic radius

Note that shielding is not important because the atoms have the same number of inner electrons.

Reasons for decrease in ionisation energy down a group

■ Increase in atomic radius
■ Increase in shielding

Reasons for nitrogen having a higher first ionisation energy than oxygen

■ Nitrogen atom — $1s^2$, $2s^2$, $2p^3$ — has a half-filled $2p$ subshell
■ Oxygen atom — $1s^2$, $2s^2$, $2p^4$ — has a more than half-filled $2p$ subshell
■ Half-filled $2p$ subshell is more stable

Successive ionisation energies

A large break occurs in successive ionisation energies of an element when moving from one energy level to another that is closer to the nucleus.

Figure 13 shows the successive ionisation energy values for sodium. A log scale is used because there is such a large difference in ionisation values that it would be impossible to plot on a conventional scale. For example, the first ionisation energy of sodium is $+500\,kJ\,mol^{-1}$ but the eleventh ionisation energy is $+158\,700\,kJ\,mol^{-1}$.

Often a question may give successive ionisation energies for some elements and ask you to determine the element in a particular group of the periodic table. Look for the large jump in successive ionisation energies. This jump occurs after all the outer shell electrons have been removed and will indicate the number of electrons in the outer shell.

Exam tip

With the half filled (group V) and filled (groups II and 0) subshells, always state the electronic configuration of the atoms involved and then explain about the stability. For example: a phosphorus atom is $1s^2$, $2s^2$, $2p^6$, $3s^2$, $3p^3$; the half-filled $3p$ subshell is more stable.

Exam tip

If you are asked to sketch a graph of successive ionisation energy values for a particular element, remember to do it from the outer electrons to the inner. For example, at GCSE sodium is 2,8,1, but the successive graph should show one electron (large gap), then eight electrons (large gap), and finally two electrons. The graph should be constantly increasing.

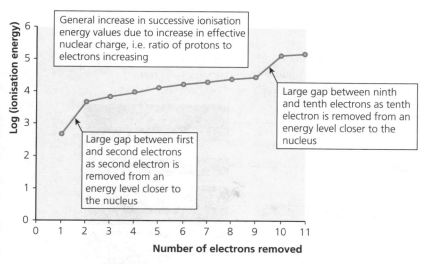

Figure 13 Successive ionisation energies of sodium

Worked example

An element has the successive ionisation energies shown in Table 6.

Table 6

Ionisation energy	First	Second	Third	Fourth	Fifth	Sixth
kJ mol^{-1}	580	1800	2700	11600	14800	18400

The large jump in successive ionisation energy values occurs after three electrons are removed, so there must be three electrons in the outer shell, which would indicate a group III element.

Summary

- Atoms are composed of protons, neutrons and electrons. Electrons have a much smaller mass than protons and neutrons. Protons have a relative charge of +1 and electrons −1. Neutrons have no charge.
- Mass spectrometry measures the mass of particles in a sample based on each having a single positive charge.
- For an element, mass spectrometry measures the mass and relative abundance of each isotope of the element.
- Elements can be classified as *s* block, *p* block or *d* block, depending on which subshell their outer shell electrons are found in.

- The first series of transition metal atoms lose their 4*s* electrons first.
- The first ionisation energy is the energy required to convert 1 mole of gaseous atoms into gaseous ions with a single positive charge.
- Ionisation energies increase across a period but decrease down a group.
- The four factors which are used to explain patterns in ionisation energies are: atomic radius, nuclear charge, shielding by inner electrons and stability of filled and half-filled subshells.

Bonding

All pure substances may be classified as elements or compounds. Figure 14 shows the main subdivisions of all pure substances. The type of bonding and structure shown by each type of substance with some common examples are also given.

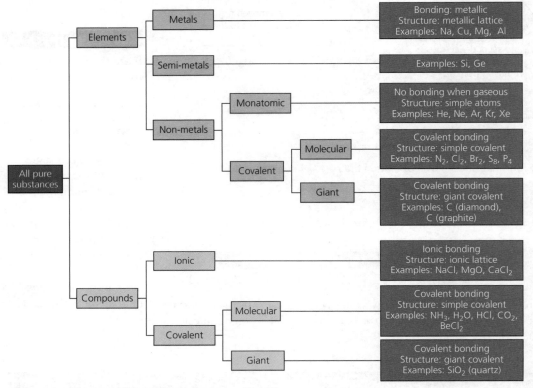

Figure 14 Types of structure and bonding

Bonding and structure are considered together in this section. Each type of structure (metals, ionic compounds, giant covalent structures and molecular covalent structures) will be examined. Semi-metals can also be called metalloids and they have properties of both metals and non-metals. Silicon has a giant covalent structure.

Metals

Metallic bonding (Figure 15) involves the attraction between layers of positive ions and **delocalised electrons**. The structure is described as a metallic lattice. A lattice is a regular arrangement of atoms or ions. The physical properties of metals are listed in Table 7.

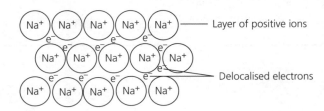

Figure 15 Typical diagram showing metallic bonding

Knowledge check 12

State the type of bonding found in calcium chloride.

Delocalised electrons are outer electrons that do not have fixed positions but move freely.

Table 7 Physical properties of metals related to bonding and structure

Physical property	Explanation of physical property in terms of structure and bonding
Hardness	Strong attraction between positive ions and negative electrons, and a regular structure
High melting point	Large amount of energy is required to break the bonds, which are strong attractions between positive ions and negative electrons
Good electrical conductivity	Delocalised electrons can move and carry charge through the metal
Malleability and ductility	Layers of positive ions can slide over each other without disrupting the bonding

The metallic bond is stronger when there are more delocalised electrons. The melting point of sodium is 98°C whereas the melting point of magnesium is 625°C. Magnesium has two electrons in its outer energy level, whereas sodium has only one. In the metallic lattice, magnesium atoms can delocalise two electrons per atom, whereas sodium atoms can only delocalise one per atom. The more electrons that are delocalised, the stronger the metallic bond. d block metals have many outer energy level electrons and so they have higher melting points than s block elements. For example, iron has a melting point of 1538°C.

Ionic compounds

Ionic bonding is the electrostatic attraction between oppositely charged ions in a regular ionic lattice. The structure of an ionic compound is described as an ionic lattice.

Ionic compounds generally contain a metal, particularly a group I or II metal, and a non-metal, particularly a group VI or VII non-metal. These simple ions have the electron configuration of a noble gas.

Positive ions are called cations and negative ions are called anions. Simple cations have the same name as the parent atom, e.g. sodium ion, Na^+, hydrogen ion, H^+, aluminium ion, Al^{3+}. Simple anions have an -ide ending, e.g. oxide, O^{2-}, chloride, Cl^-, hydride, H^-, nitride, N^{3-}.

Other common simple cations found in ionic compounds include silver(I), Ag^+; copper(II), Cu^{2+}; iron(II), Fe^{2+}; iron(III), Fe^{3+}; nickel(II), Ni^{2+}; cobalt(II), Co^{2+}; zinc, Zn^{2+}; manganese(II), Mn^{2+}.

Most simple cations formed from d block elements do not have a noble gas electron configuration.

Positive molecular ions end in -onium. The most common molecular ion is the ammonium ion, but you will also come across the hydroxonium ion, H_3O^+.

Negative molecular ions usually end in -ate. However, some common names end in -ite.

Other common molecular anions found in ionic compounds include sulfate, SO_4^{2-}; carbonate, CO_3^{2-}; nitrate, NO_3^-; hydroxide, OH^-; hypochlorite, OCl^-; hydrogencarbonate, HCO_3^-; dichromate or dichromate(VI), $Cr_2O_7^{2-}$; permanganate or manganate(VII), MnO_4^-.

Dot-and-cross diagrams

Ionic compounds are formed when metal atoms transfer electrons to non-metals atoms. A dot-and-cross diagram can be used to show how electrons are transferred:

Knowledge check 13

What is meant by metallic bonding?

Exam tip

There are a few negative molecular ions that end in -ide. Examples are hydroxide (OH^-) and cyanide (CN^-). Remember these — they are unusual.

Exam tip

Remember that the formula of an ionic compound can be worked out using the charges on the ions: e.g. ammonium carbonate contains NH_4^+ and CO_3^{2-} ions. Two ammonium ions are required to cancel out the charge on the carbonate ion, so the formula of ammonium carbonate is $(NH_4)_2CO_3$.

- Only outer shell electrons are shown on all atoms and ions.
- The correct number of each atom required must be shown, together with the correct number of each ion in the compound.
- Ions should be placed in square brackets with the transferred electrons shown in the ions (using × or •) and the charge on all the ions should be shown outside the brackets. Figure 16 shows the dot-and-cross-diagram for magnesium chloride.

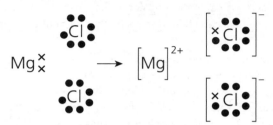

Figure 16 Magnesium chloride

Figure 17 shows a second example, for calcium oxide.

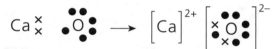

Figure 17 Calcium oxide

Properties of ionic compounds are given in Table 8.

Table 8 Properties of ionic compounds related to bonding and structure

Physical property	Explanation of physical property in terms of structure and bonding
Crystalline	Regular lattice of positive and negative ions. Regular arrangement creates crystal structure
High melting point and boiling point	Large amount of energy is required to break the bonds which are strong electrostatic attractions between ions of opposite charge
Non-conductor of electricity when solid	Ions are not free to move and cannot carry charge
Good conductor of electricity when molten or when aqueous (dissolved in water)	Ions are free to move and can carry charge

Sodium chloride

Sodium chloride has a 6:6 crystal arrangement, which means that six Na^+ ions surround one Cl^- ion and six Cl^- ions surround one Na^+ ion. Figure 18 shows part of the structure.

Melting point and boiling point

The energy required to melt an ionic compound is largely due to the number of strong electrostatic attractions between the positive and negative ions. The smaller the ions

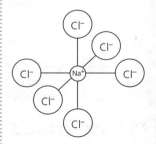

Figure 18 Part of an ionic lattice

and the higher their charge, the stronger the ionic bond. Magnesium oxide has a melting point of 2852°C, whereas sodium chloride has a melting point of 797°C. The attraction between the Mg^{2+} and O^{2-} ions is stronger than the attraction between the Na^+ and Cl^- ions. Some atomic and ionic radii measured in picometres (pm) are given in Figure 19. Metal cations are generally smaller than the atoms from which they are formed while non-metal anions are generally larger than the atoms from which they are formed.

Exam tip

1 pm = 10^{-12} m. There are 1000000000000 pm in 1 m. Sometimes nanometres (nm) are used. 1 nm = 10^{-9} m. There are 1000 pm in 1 nm. It is important to be able to recognise units of this scale when dealing with the size of particles and ions.

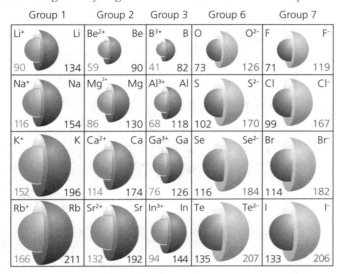

Figure 19 Sizes of atoms and simple ions. Numbers in black are atomic radii, numbers in red are ionic radii for cations, numbers in blue are ionic radii for anions (all numbers are pm)

Covalent bonding

A single **covalent bond** is a shared pair of electrons, represented as a line between two atoms, e.g. H–Cl. A double covalent bond is two shared pairs of electrons, represented as a double line between two atoms, e.g. O=C=O. A triple covalent bond is three shared pairs of electrons, represented as a triple line between two atoms, e.g. N≡N.

A **covalent bond** is the electrostatic attraction between a shared pair of electrons and the nuclei of bonded atoms.

Covalent bonds exist between non-metal atoms (some exceptions are where metal atoms can form covalent bonds, e.g. Be in $BeCl_2$ and Al in $AlCl_3$).

Dot-and-cross diagrams are again used to show the arrangement of electrons in covalently bonded molecules. A shared pair of electrons may be represented as ×● to show that the two electrons in the bond are from different atoms (Figure 20).

Knowledge check 14

What is meant by a covalent bond?

The **octet rule** states that when forming a compound, an atom tends to gain, lose or share electrons to achieve eight electrons in its outer shell.

Hydrogen is not considered as part of the octet rule because it can only have a maximum of two electrons in its outer shell.

Some molecules have atoms that deviate from the octet rule. They are said to have contracted their octet (i.e. have fewer than eight electrons in their outer shell) or expanded their octet (i.e. have more than eight electrons in their outer shell) — see Figure 21.

The **octet rule** states that atoms tend to gain, lose or share electrons when they combine with other atoms to achieve a stable octet of electrons.

Dot-and-cross diagrams for common molecules with single covalent bonds

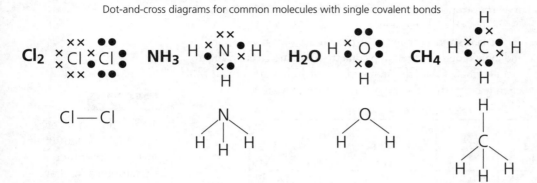

Dot-and-cross diagrams for common molecules with multiple covalent bonds

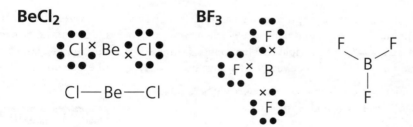

Figure 20 Dot-and-cross diagrams for covalent bonding

Common examples of contraction of octet

BeCl₂

BF₃

Common example of expansion of octet

SF₆

Figure 21 Deviation from the octet rule

Bonding pairs and lone pairs of electrons

Methane has four **bonding pairs** of electrons. An ammonia molecule has three bonding pairs of electrons and one **lone pair** of electrons (Figure 22). The number of bonding pairs of electrons and lone pairs of electrons helps to determine the shape of a molecule but is also important for the formation of a coordinate bond.

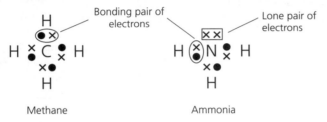

Methane   Ammonia

Figure 22 Bonding pairs and lone pairs of electrons

Coordinate bonds

Coordinate bonds (also called dative covalent bonds) form when one atom contributes both of the shared pair of electrons. When a coordinate bond forms a lone pair of electrons becomes a bonding pair of electrons. There are many examples of coordinate bonded ions.

Note that in a coordinate bonded ion the charge must be shown — see for example the ammonium ion (Figure 23).

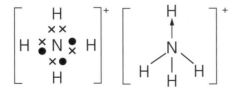

Figure 23 The ammonium ion

In the dot-and-cross diagram, a coordinate bond is shown with two crosses to show that the two electrons come from the same atom. It is also shown as an arrow in the bonding diagram, with the direction indicating where the two electrons come from. Note that once a coordinate bond is formed it is exactly the same as a covalent bond.

Electronegativity and polarity

In a covalent bond the two atoms at either end of the bond have an **electronegativity** value. Electronegativity is the numerical value of the ability of an atom to attract the bonding electrons in a covalent bond.

For example, in H–Cl the electronegativity values are H = 2.1 and Cl = 3.0. This means that the electrons in this bond are drawn closer to Cl than to H because Cl has a higher electronegativity value. This is represented by the use of partial charges, represented by $\delta+$ (delta plus) and $\delta-$ (delta minus) above the atoms in the bond. The $\delta-$ is placed above the atom that has the higher electronegativity value and the $\delta+$ is placed above the atom with the lower electronegativity value (Figure 24). This covalent bond is now described as being polar.

$$\overset{\delta+}{H} - \overset{\delta-}{Cl}$$

Figure 24 Polarity of H–Cl bond

A **polar bond** is one where the atoms at either end of the bond have different electronegativity values, resulting in an unequal sharing of the bonding electrons (Figure 25).

Examples of polar bonds: $\overset{\delta-}{O}-\overset{\delta+}{H}$ $\overset{\delta-}{N}-\overset{\delta+}{H}$ $\overset{\delta+}{H}-\overset{\delta-}{F}$ $\overset{\delta+}{C}=\overset{\delta-}{O}$

Examples of non-polar bonds: I–I   C–H   C=C   Cl–Cl

Figure 25 Polar and non-polar bonds

Trends in electronegativity

- As a group is descended, electronegativity decreases (bonded electrons are further from the attractive power of the nucleus).
- As a period is crossed from left to right, electronegativity increases (as bonded electrons are closer to the attractive power of the nucleus).

The most electronegative element is fluorine. The least electronegative element is caesium.

Ionic and covalent character

For compounds composed of two different elements, the difference in electronegativity between the atoms of the elements dictates the type of compound formed (i.e. ionic or covalent) and, if the compound is covalent, the polarity of the molecule.

- No/very small difference in electronegativity gives a non-polar molecule, e.g. Br_2, I_2, Cl_2, O_2, CH_4
- Small difference in electronegativity gives a polar molecule, e.g. HF, HCl, H_2O, NH_3
- Large difference in electronegativity gives an ionic compound, e.g. NaCl, MgO, CaF_2

Simple covalent molecules can be polar or non-polar depending on whether or not they contain polar bonds but also based on their shape.

Electronegativity is the extent to which an atom attracts the bonding electrons in a covalent bond.

A **polar bond** is a covalent bond in which there is unequal sharing of the bonding electrons.

Exam tip

Most non-polar covalent bonds have the same atom at either end of the bond but some have only a very small difference in electronegativity, e.g. C–H.

Exam tip

Francium should be more electronegative than caesium but francium is radioactive so it is unstable and its chemistry is not considered because it cannot be isolated in sufficient quantity. This is also true of other elements in period 6, such as polonium, astatine and radon, and elements in period 7.

Knowledge check 16

Explain which is more electronegative, chlorine or iodine.

If a molecule contains equally polar bonds arranged symmetrically then the polarity of the bonds cancel each other and the molecule is non-polar — for example, carbon dioxide, CO_2, is a linear molecule (Figure 26).

$$\overset{\delta-}{O} = \overset{\delta+}{C} = \overset{\delta-}{O}$$

Figure 26 Equally polar bonds arranged symmetrically leads to no overall dipole

If equally polar bonds are arranged symmetrically, the bond polarities cancel each other out and the molecule has no overall dipole. Symmetrical shapes are linear, trigonal planar, tetrahedral and octahedral.

Summary

- Substances are either pure or mixtures.
- Pure substances are either elements or compounds; elements are either metals or non-metals (or semi-metals).
- Metals exhibit metallic bonding (positive ions held together by a field of delocalised electrons).
- Most compounds containing a metal exhibit ionic bonding (attractions between oppositely charged ions); exceptions are $BeCl_2$ and $AlCl_3$.
- Compounds containing only non-metals exhibit covalent bonding (sharing of electrons).
- Some non-metallic elements and compounds form simple molecules that exhibit covalent bonding.
- Electronegativity is the extent to which an atom attracts the bonding electrons in a covalent bond.
- Simple covalent molecules can be described as polar or non-polar based on the difference in electronegativity of the atoms involved and also the symmetry of the molecule.

■Intermolecular forces

Intermolecular forces are the bonds that exist between neighbouring simple covalent molecules. There are three types of intermolecular forces.

Van der Waals forces are attractions between induced dipoles (temporary dipoles caused by random movement of electrons around atoms). Van der Waals forces exist between all simple molecules and atoms in the liquid and solid states. Remember there are no forces of attraction between molecules in a gas.

Van der Waals forces are the only forces of attraction between non-polar molecules such as iodine (I_2), bromine (Br_2), sulfur (S_8), carbon dioxide (CO_2) in the solid state and liquid tetrachloromethane (CCl_4).

The more electrons are present in a molecule the greater the van der Waals forces of attraction — this explains the increase in boiling point as the chain increases in length in alkanes and the increase the boiling point on going down group VII. Often RMM (or molar mass) is used as a measure of the number of electrons in a molecule. Molecules with similar RMMs have comparable van der Waals forces of attraction.

Van der Waals forces are attractions between instantaneous and induced dipoles on neighbouring molecules.

Permanent dipole–dipole attractions

When a simple covalent molecule is polar, it is said to have a **permanent dipole**. The forces of attraction between polar molecules are van der Waals forces *and* **permanent dipole–dipole attractions**.

Permanent dipole–dipole attractions exist between polar molecules like propanone (CH_3COCH_3) and trichloromethane $(CHCl_3)$ — see Figure 27.

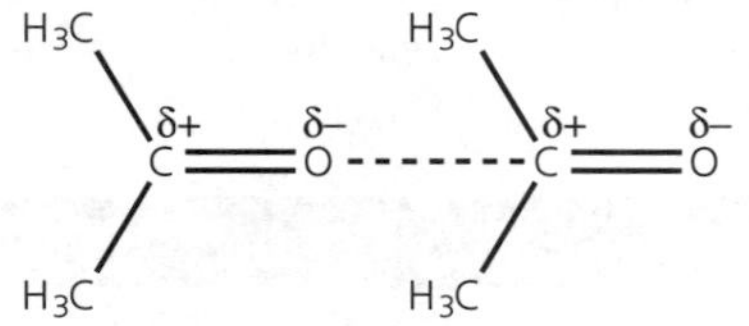

Figure 27 Permanent dipole–dipole attraction

The permanent dipole–dipole attraction is the attraction between the $\delta+$ on one molecule and the $\delta-$ on another molecule. It is important to label the polarities $(\delta+$ and $\delta-)$ of the polar bonds involved.

Hydrogen bonds (H bonds) are intermolecular forces where the bond is formed between a $\delta+$ H atom and a $\delta-$ N, O or F atom of another molecule. The hydrogen bond is formed because of the attraction between a lone pair of electrons on the $\delta-$ atom and the $\delta+$ hydrogen atom (Figure 28). Hydrogen bonds are often shown as dashed lines between the H (bonded to N, O or F) and the $\delta-$ atom on another molecule.

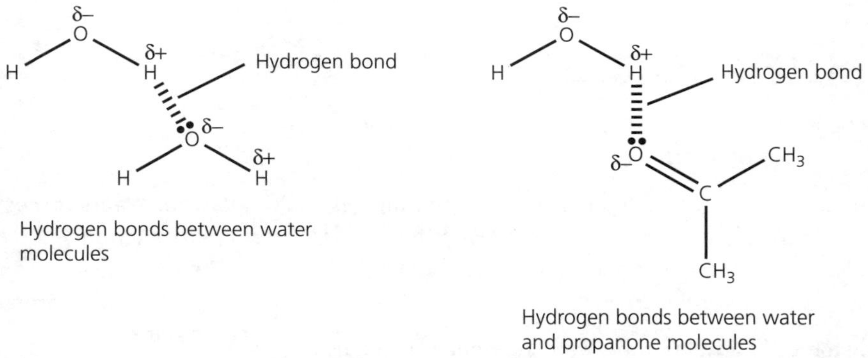

Hydrogen bonds between water molecules

Hydrogen bonds between water and propanone molecules

Figure 28 Hydrogen bonds

Explanation of properties using intermolecular forces

Intermolecular forces are used to explain a physical property — for example, melting point or boiling point — of a simple covalent substance. For example, 'a large amount of energy is needed to break the [intermolecular force(s) — state which one(s)] between the molecules.

They are also used when explaining viscosity, which is the opposite of fluidity (the more viscous a liquid the less well it flows). For example, 'the greater the [intermolecular force(s) — state which one(s)] between the molecules, the greater the viscosity.

Permanent dipole–dipole attractions are the attractions between the positive end, $\delta+$, of the permanent dipole on a molecule with the negative end, $\delta-$, of the permanent dipole of a neighbouring molecule.

A **hydrogen bond** is the attraction between a lone pair of electrons on a very electronegative atom (i.e. N, O, F) in one molecule and a hydrogen atom in a neighbouring molecule in which the hydrogen atom is covalently bonded to a very electronegative atom (N, O, F).

Exam tip

When drawing a diagram like this, make sure you label the polarities as $\delta+$ of the H atom and $\delta-$ of the other atom, and show the hydrogen bond as a dashed line from the lone pair and label the line 'hydrogen bond'.

Knowledge check 17

Name two types of intermolecular force.

Liquids can be described as miscible (able to mix together in all proportions, forming one layer) or immiscible (unable to mix together and forming two distinct layers). Liquids often mix with each other because of their ability to form the same intermolecular forces between the molecules. Solids are soluble in a solvent because they have similar bonds between their molecules.

Worked example 1

Explain why water has a higher than expected boiling point.

Answer

A large amount of energy is needed to break the hydrogen bonds between water molecules.

Worked example 2

Explain why iodine has a higher boiling point than bromine.

Answer

I_2 has more electrons than Br_2 so there are greater van der Waals forces of attraction between the I_2 molecules.

Worked example 3

Explain why the liquid alkanes increase in viscosity as the carbon chain increases in length.

Answer

The number of electrons increases as the carbon chain length increases so van der Waals forces are stronger between molecules, making it less fluid in the liquid state.

Worked example 4

Explain why ethanol mixes with water.

Answer

Ethanol has OH groups that can form hydrogen bonds with water molecules.

> **Exam tip**
>
> 'Like dissolves like' is a phrase often used to explain miscibility of liquids and the solubility of a solid in a solvent. 'Like' in this case means that the bonding between the 'particles' in one substance is similar to the bonding between the 'particles' in the other substance.

Worked example 5

Explain why bromine mixes with hexane.

Answer

Bromine and hexane are both non-polar and like dissolves like.

Worked example 6

Explain why sodium chloride dissolves in water.

Answer

Sodium chloride is an ionic substance and water is polar — like dissolves like.

Hydrides of groups IV, V, VI and VII

Figure 29 shows the boiling points of the hydrides of groups IV, V, VI and VII.

The boiling points of H_2O, HF and NH_3 are higher than expected. This is because of hydrogen bonds between the molecules of these compounds. CH_4 does not form hydrogen bonds between its molecules because it is non-polar.

Note in Figure 29 the increase in boiling point from H_2S to H_2Te (and the same in the other groups). This increase is due to the higher number of electrons in the molecules, which increases the van der Waals forces of attraction between the molecules.

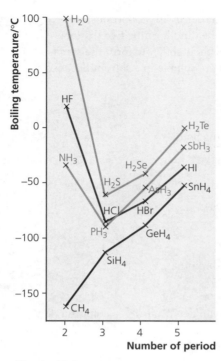

Figure 29 Boiling points of hydrides

Exam tip

You must be able to explain the higher than expected boiling points of the hydrides of the first elements in groups V, VI and VII, and the reason why CH_4 has a boiling point that fits the pattern for group IV — CH_4 is non-polar so does not form hydrogen bonds. You must also be able to explain the pattern in increasing boiling point going down the groups in terms of van der Waals forces of attraction.

Ice

Ice has a lower density than water, so ice floats on water. This is because the hydrogen bonds in ice are more ordered (fixed) and the water molecules in ice are further apart, leading to a more open structure and so a lower density.

Figure 30 shows the arrangement of water molecules in water and ice. There is more space between the water molecules in ice compared to liquid water.

Knowledge check 18

Explain why ice has a lower density than liquid water.

Figure 30 Water (left) and ice (right)

Summary

- Intermolecular forces exist between simple molecules.
- Simple molecules can be molecules of elements or compounds.
- The three forces of attraction between simple molecules are van der Waals forces of attraction, permanent dipole–dipole attractions and hydrogen bonds.
- Non-polar molecules and simple atoms (like the noble gases) have only van der Waals forces of attraction between their particles in the liquid and solid states.
- Van der Waals forces of attraction are caused by induced dipoles, which are the result of the random movement of electrons in the molecule.
- The greater the number of electrons in a molecule (or atom), the greater the van der Waals forces of attraction.
- Polar molecules have permanent dipole–dipole attractions between the molecules, as well as van der Waals forces of attraction.
- Hydrogen bonds occur between molecules where there is a H atom bonded to a N, O or F atom which bonds with a lone pair of electrons on a N, O or F atom of another molecule.
- Hydrogen bonds are the strongest of the intermolecular forces of attraction.
- Physical properties of simple covalent elements or compounds can be explained in terms of the intermolecular forces between their molecules.

Structure

There are four main types of structure — ionic, metallic, giant covalent and molecular covalent.

The properties of ionic and metallic substances have already been considered (pp. 30–33). Covalent substances are either molecular covalent or giant covalent. Many solid molecular covalent substances are molecular covalent crystals.

Molecular covalent crystals

The forces of attraction between the molecules in molecular covalent substances are usually so weak that they are gases, liquids or low melting point solids. When they are solid, many form molecular covalent crystals. The main examples of molecular covalent crystals are iodine and sulfur. Ice is another example of a molecular covalent crystal.

Sulfur exists as S_8 molecules, which consist of eight sulfur atoms covalently bonded together in a puckered ring. Iodine exists as a simple diatomic molecule, I_2 (Figure 31).

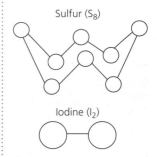

Sulfur (S_8)

Iodine (I_2)

Figure 31 Sulfur and iodine

Properties of molecular covalent crystals linked to bonding and structure

Molecular covalent crystals usually have low melting/boiling points due to weak intermolecular forces. They do not conduct electricity in any state because they have no free electrons/ions to carry charge.

Giant covalent crystals

The two main examples of giant covalent (or macromolecular) crystals are carbon (graphite) and carbon (diamond). The strong covalent bonds between the atoms in these crystals mean that they are high melting point solids (a lot of energy is needed to break the large number of strong covalent bonds). The regular structure means that they are crystalline.

Content Guidance

In diamond each carbon atom is covalently bonded to four others in a tetrahedral arrangement (Figure 32a). The rigid three-dimensional structure of diamond combined with the strong covalent bonds means that it is hard.

In graphite each carbon is covalently bonded to three others in a layered hexagonal structure (Figure 32b). The delocalised electrons between layers can move and carry charge, so in the solid state graphite can conduct electricity. Molten graphite does not conduct electricity because the structure is disrupted — whereas molten metals continue to conduct electricity. Diamond does not conduct electricity because there are no delocalised electrons to move and carry the charge.

Graphite can act as a lubricant since the layers can slide over each other due to the weak bonds between them.

Knowledge check 19

Explain why graphite conducts electricity.

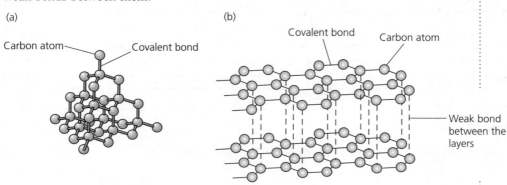

Figure 32 The structure of (a) diamond and (b) graphite

Table 9 summarises the properties of the different types of crystalline substances.

Table 9

	Metals	Ionic compounds	Molecular (simple) covalent	Macromolecular (giant) covalent
Common examples	Magnesium	Sodium chloride	Ice, iodine	Diamond, graphite
Bonding	Metallic	Ionic	Covalent within the molecules and intermolecular forces between the molecules	Covalent (graphite has weak bonds between the layers in the structure)
Electrical conductivity	Conduct electricity when solid and molten	Do not conduct electricity when solid Conduct electricity when molten or dissolved in water	Does not conduct electricity	Does not conduct electricity (graphite does conduct electricity as a solid)
Melting point	Generally high melting points	High melting points	Low melting points	High melting points
Solubility in water	Insoluble in water (some metals will react with water)	Generally soluble in water	Mostly insoluble in water (some polar substances dissolve in water and some non-polar substances react with water)	Insoluble in water

Periodic trends related to structure and bonding

From sodium to argon, the melting point increases to silicon then decreases.

- From Na to Mg to Al the metallic bond increases in strength because there are more outer shell electrons that can be delocalised, so a greater attraction between the electrons and the ions in the metallic structure.
- Silicon has a giant covalent structure and so has the highest melting point in the period because a substantial amount of energy is required to break the large number of strong covalent bonds.
- Phosphorus (P_4), sulfur (S_8) and chlorine (Cl_2) are non-polar simple covalent molecules with low melting points. There are only van der Waals forces of attraction between the molecules in these three substances. S_8 has the most electrons and so the greatest van der Waals forces of attraction between molecules. Argon is monatomic.

Summary

- There are four types of crystalline structure: metallic, ionic, molecular covalent and giant covalent.
- The physical properties of these crystalline substances depend on the strength of the bonds within the crystal.
- Metallic crystals generally have high melting points and conduct electricity.
- Ionic crystals generally have high melting points, are soluble in water and conduct electricity when molten or aqueous.
- Molecular covalent crystals generally have low melting points, are insoluble in water and do not conduct electricity.
- Giant covalent crystals generally have high melting points, are insoluble in water and do not conduct electricity (graphite does conduct electricity).

■ Shapes of molecules and ions

The shape of a covalent molecule or an ion depends on the repulsion of the electrons around a central atom. The electron pairs are charge clouds around an atom and they repel each other as far as possible. There are two types of electron pair, a **bonding pair** of electrons and a **lone pair** (non-bonding pair) of electrons (Figure 33).

The shape of the molecule or ion is determined from:
- the total number of electron pairs around a central atom
- the number of bonding pairs of electrons
- the number of lone pairs of electrons

You must be able to identify the lone pairs and bonding pairs of electrons in any molecule or ion. Lone pairs are more compact to the central atom so they have a greater repulsive effect on the other pairs of electrons.

The order of strength of the repulsions experienced by the electron pairs is shown in Figure 34.

A **bonding pair** of electrons is a pair of electrons shared between two atoms.

A **lone pair** of electrons is a pair of unshared electrons in the outer shell of an atom.

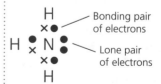

Figure 33 Example of a lone pair of electrons and a bonding pair of electrons

Content Guidance

| Lone pair ⟷ Lone pair **LP ⟷ LP** | is greater than | Lone pair ⟷ Bonding pair **LP ⟷ BP** | is greater than | Bonding pair ⟷ Bonding pair **BP ⟷ BP** |

Figure 34 The order of strength of repulsions

This means that lone pairs of electrons repel lone pairs of electrons more than they repel bonding pairs of electrons. The lowest level of repulsion is between bonding pairs of electrons. The molecule or ion will take up a shape that minimises these repulsions. The shape depends on the arrangement of atoms around the central atom.

In questions relating to shapes of molecules and ions you may be asked for any combination of the following:
1 a sketch of the shape
2 the name of the shape
3 the bond angle
4 an explanation of the shape

For the examples that follow, all of the above are given for each molecule or ion.

Examples with only bonding pairs of electrons

Beryllium chloride ($BeCl_2$) is shown in Figure 35. There are two bonding pairs of electrons around the beryllium atom. These repel each other equally so the molecule takes up a **linear** shape to minimise the effect of the repulsions. The bond angle is the angle between the two covalent bonds, which in beryllium chloride is 180°.

Sketch of the shape

Bond angle = 180°
Shape = Linear
Explanation: two bonding pairs of electrons repel each other equally and the molecule takes up this shape to minimise repulsions

Figure 35 Beryllium chloride

Boron trifluoride (BF_3) is shown in Figure 36. There are three bonding pairs of electrons around the boron atom in BF_3. These repel each other equally and so the molecule takes up a **trigonal planar** shape with a bond angle of 120°.

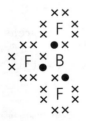 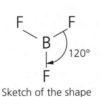

Sketch of the shape

Bond angle = 120°
Shape = Trigonal planar
Explanation: three bonding pairs of electrons repel each other equally and the molecule takes up this shape to minimise repulsions

Figure 36 Boron trifluoride

Methane (CH_4) is shown in Figure 37. There are four bonding pairs of electrons around the carbon atom in methane. These repel each other equally and the molecule takes up a **tetrahedral** shape to minimise repulsions. The bond angle is 109.5°.

The term tetrahedral refers to the solid shape formed when all the hydrogen atoms are connected — a triangular based pyramid with four sides called a tetrahedron.

Exam tip

Linear and trigonal planar shapes can be drawn easily as they are two-dimensional. When there are four or more pairs of electrons the arrangement becomes three-dimensional, which requires a little more skill in drawing.

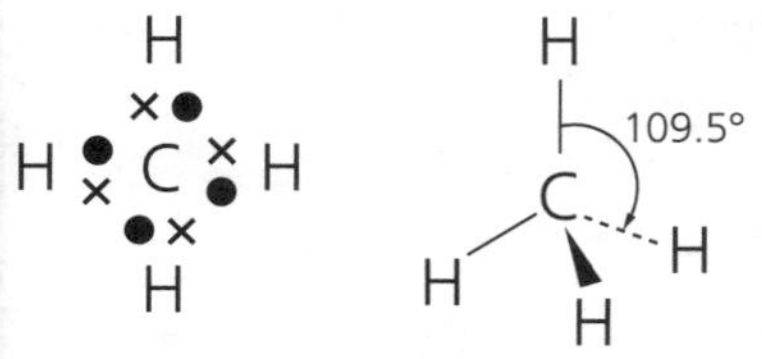

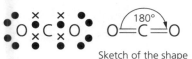

Sketch of the shape

Figure 37 Methane

Bond angle = 109.5°
Shape = Tetrahedral
Explanation: four bonding pairs of electrons repel each other equally and the molecule takes up this shape to minimise repulsions

Drawing three-dimensional shapes

When drawing a three-dimensional shape such as the tetrahedral shape of methane, three types of lines are drawn (see Figure 38) to show the three-dimensional arrangement of the atoms. Bonds in the plane of the page are shown as normal lines (—). Bonds coming towards the viewer out of the plane of the page are drawn using a solid wedge getting thicker as it comes out towards the atom at the end of the bond (►). Bonds going backwards from the plane of the paper are shown using a dashed line (----).

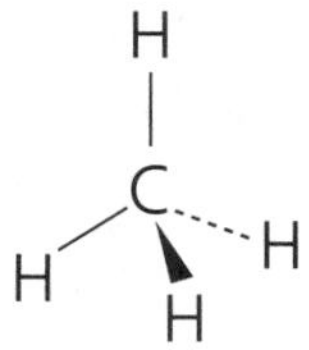

Carbon dioxide (CO_2) is shown in Figure 39. The carbon in carbon dioxide has two sets of bonding pairs of electrons. A double bonding pair of electrons repels in the same way as a single bonding pair. The two sets of bonding pairs of electrons repel each other equally so CO_2 takes up a linear shape to minimise repulsions.

Figure 38

O=C=O   180°

Sketch of the shape

Bond angle = 180°
Shape = Linear
Explanation: two (sets of) bonding pairs of electrons repel each other equally and the molecule takes up this shape to minimise repulsions

Figure 39 Carbon dioxide

A double bonding pair of electrons or a triple bonding pair of electrons repel in the same way as a single bonding pair. This can be seen from the bonding diagram in Figure 40 for hydrogen cyanide (HCN) and ethene (C_2H_4).

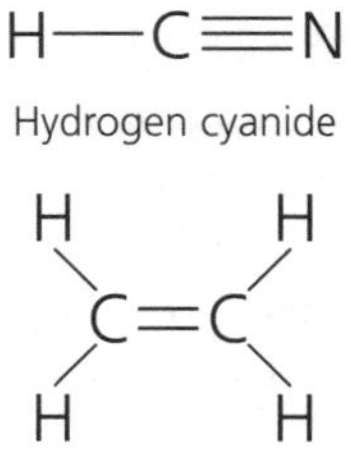

Hydrogen cyanide is linear due to equal repulsions of the triple bonding pair of electrons and the single bonding pair of electrons. Around the carbon atoms in ethene the shape is trigonal planar due to the equal repulsion of the three sets of bonding pairs of electrons (even though one is a double set).

Phosphorus pentafluoride (PF_5) is shown in Figure 41. There are five bonding pairs of electrons around the central phosphorus atom. These bonding pairs of electrons repel each other equally and the molecule takes up a **trigonal bipyramidal** shape. There are two bond angles in a trigonal bipyramid, 90° and 120°.

H—C≡N
Hydrogen cyanide

Ethene

Figure 40 Hydrogen cyanide and ethene

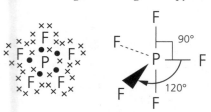

Sketch of the shape

Bond angle = 90° and 120°
Shape = Trigonal bipyramidal
Explanation: five bonding pairs of electrons repel each other equally and the molecule takes up this shape to minimise repulsions

Figure 41 Phosphorus pentafluoride

Again if the points where the fluorine atoms are placed are connected, the shape formed is a triangle with a pyramid above and below. This is called a trigonal bipyramid.

Sulfur hexafluoride (SF_6) is shown in Figure 42. There are six bonding pairs of electrons around the central sulfur atom. These repel each other equally and the molecule takes up an **octahedral** shape to minimise the repulsions.

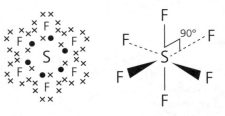

Bond angle = 90°
Shape = Octahedral
Explanation: six bonding pairs of electrons repel each other equally and the molecule takes up this shape to minimise repulsions

Sketch of the shape

Figure 42 Sulfur hexafluoride

Knowledge check 21

State the shape and bond angle in sulfur hexafluoride.

The octahedral shape is also called square bipyramidal. The term octahedral comes from the fact that the solid shape formed from connecting all the fluorine atoms forms an eight-sided figure called an octahedron.

Examples with bonding pairs of electrons and lone pairs of electrons

The following examples all have four pairs of electrons around the central atom. These pairs of electrons take up a tetrahedral shape like CH_4. However, out of the four pairs of electrons, some are bonding pairs of electrons and some are lone pairs of electrons. Remember, a lone pair of electrons has a greater repulsion than a bonding pair of electrons.

Ammonia (NH_3) is shown in Figure 43. There are three bonding pairs of electrons and one lone pair of electrons around the central nitrogen atom in NH_3. The basic arrangement of the electron pairs is tetrahedral around the nitrogen but since there is no atom attached to the lone pair, all you see is the bottom of the tetrahedron, which looks like a pyramid. The extra repulsion from the lone pair squeezes the bonding pairs of electrons closer together, decreasing the bond angle to 107°.

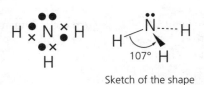

Bond angle = 107°
Shape = Pyramidal
Explanation: three bonding pairs of electrons and one lone pair of electrons; the lone pair of electrons has a greater repulsion than the bonding pairs of electrons and the molecule takes up this shape to minimise repulsions

Sketch of the shape

Figure 43 Ammonia

Water (H_2O) is shown in Figure 44. The basic arrangement of the electron pairs is tetrahedral around the oxygen atom but as there is no atom attached to the lone pairs all you see is two bonds of the tetrahedron, which appears bent. The extra repulsion from the lone pairs squeezes the bonds closer giving a **bent** (or V) shape and decreasing the bond angle to 104.5°.

Bond angle = 104.5°
Shape = Bent
Explanation: two bonding pairs of electrons
and two lone pairs of electrons; the lone pairs
of electrons have a greater repulsion than the
bonding pairs of electrons; the molecule takes
up this shape to minimise repulsions

104.5°

Sketch of the shape

Figure 44 Water

Examples involving coordinate bonds

When a coordinate bond forms it converts a lone pair of electrons into a bonding pair of electrons.

Ammonia reacts with hydrogen ions to form the ammonium ion, NH_4^+. The formation of the coordinate bond causes a change in the shape. Remember, ammonia (NH_3) is pyramidal (three bonding pairs of electrons and one lone pair of electrons) but the ammonium ion is tetrahedral (four bonding pairs of electrons).

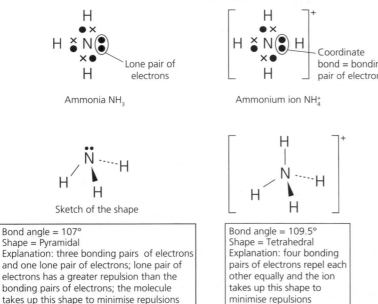

Lone pair of electrons

Ammonia NH_3

Coordinate bond = bonding pair of electrons

Ammonium ion NH_4^+

Sketch of the shape

Bond angle = 107°
Shape = Pyramidal
Explanation: three bonding pairs of electrons
and one lone pair of electrons; lone pair of
electrons has a greater repulsion than the
bonding pairs of electrons; the molecule
takes up this shape to minimise repulsions

Bond angle = 109.5°
Shape = Tetrahedral
Explanation: four bonding
pairs of electrons repel each
other equally and the ion
takes up this shape to
minimise repulsions

Figure 45 Ammonia and the ammonium ion

The H_3O^+ ion is formed when H_2O reacts with H^+. The H_3O^+ ion has three bonding pairs of electrons and one lone pair of electrons around the oxygen atom, so it takes up a pyramidal shape (bond angle 107°) to minimise repulsions.

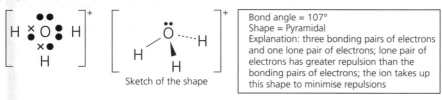

Sketch of the shape

Bond angle = 107°
Shape = Pyramidal
Explanation: three bonding pairs of electrons
and one lone pair of electrons; lone pair of
electrons has greater repulsion than the
bonding pairs of electrons; the ion takes up
this shape to minimise repulsions

Figure 46 Water and the hydronium ion

The ion BF_4^- is formed when BF_3 reacts with F^-. The ion has four bonding pairs of electrons around the central boron atom so it takes up a tetrahedral shape (bond angle 109.5°) to minimise repulsions.

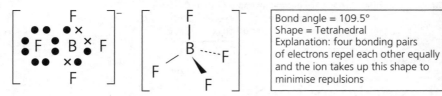

Bond angle = 109.5°
Shape = Tetrahedral
Explanation: four bonding pairs of electrons repel each other equally and the ion takes up this shape to minimise repulsions

Figure 47 BF_4^-

More unusual examples

With some more complex molecules and ions it is important to be able to visualise the total number of electrons around the central atom. This helps with the basic shape.

The number of lone pairs of electrons and bonding pairs of electrons dictate the shape and the bond angle. The lone pairs will also push the bonding pairs of electrons closer together. Each lone pair typically reduces the bond angle by around 2 to 2.5°.

Bromine trifluoride (BrF_3) is shown in Figure 48. Bromine (outer electronic configuration $4s^2 4p^5$) has one unpaired electron, so to form the three covalent bonds required for BrF_3, one electron in the bromine atom is promoted to a higher subshell. This gives bromine in this compound three unpaired electrons, which can form three bonding pairs of electrons leaving two lone pairs of electrons.

Five pairs of electrons would suggest a trigonal bipyramidal general shape with two of the pairs being lone pairs of electrons. The dots in the dot-and-cross diagram represent the bromine electrons.

The shape is described as T shaped because the lone pairs of electrons take up positions 120° from each other. The three fluorine atoms take up the three other positions in the trigonal bipyramid.

The repulsion from the lone pairs of electrons is greater than the repulsion from the bonding pairs of electrons so the 90° angle is reduced to 86°.

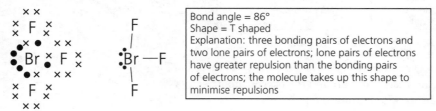

Bond angle = 86°
Shape = T shaped
Explanation: three bonding pairs of electrons and two lone pairs of electrons; lone pairs of electrons have greater repulsion than the bonding pairs of electrons; the molecule takes up this shape to minimise repulsions

Figure 48 BrF_3

Worked example 1

BrF_3 reacts to form BrF_2^+ and BrF_4^-. Determine the shape of both BrF_2^+ and BrF_4^-.

Answer

In BrF_2^+, BrF_3 has lost an F^-. BrF_3 contains two lone pairs of electrons and three bonding pairs of electrons.

BrF_2^+ will contain two lone pairs of electrons and two bonding pairs of electrons. With four pairs of electrons the basic shape is tetrahedral but, like water, the shape will be bent with a bond angle of 104.5°.

Bond angle = 104.5°
Shape = Bent
Explanation: two bonding pairs of electrons and two lone pairs of electrons; the lone pairs of electrons have a greater repulsion than the bonding pairs of electrons; the ion takes up this shape to minimise repulsions

Figure 49 BrF_2^+

In BrF_4^-, BrF_3 has gained a F^-, so has two lone pairs of electrons but four bonding pairs of electrons. This gives six electron pairs and a basic shape of octahedral where the atoms and lone pairs will be at 90° to each other.

To get as far away from each other as possible the lone pairs of electrons take up positions above and below the Br atom. This leaves the four F atoms in the ion arranged in a square planar arrangement with a bond angle of 90°.

Bond angle = 90°
Shape = Square planar
Explanation: four bonding pairs of electrons and two lone pairs of electrons; the lone pairs of electrons have a greater repulsion than the bonding pairs of electrons; the ion takes up this shape to minimise repulsions

Figure 50 BrF_4^-

Worked example 2

Determine the shape of XeF_2.

Answer

Xe promotes one electron to allow it to have two unpaired electrons and form two covalent bonds. Xe will have three lone pairs of electrons and two bonding pairs of electrons. The basic shape is trigonal bipyramidal.

The three lone pairs take up position as far away from each other as possible (120° from each other). The bonding pairs take up position above and below the Xe atom so the molecule is linear with a bond angle of 180°.

→

Knowledge check 22

Determine the shape of the PCl_4^+ and PCl_6^- ions.

Bond angle = 180°
Shape = Linear
Explanation: two bonding pairs of electrons and three lone pairs of electrons; lone pairs of electrons have a greater repulsion than the bonding pairs of electrons; the molecule takes up this shape to minimise repulsions

Figure 51 XeF_2

Table 10 summarises the shapes that you should learn.

Table 10 Summary table for shapes of molecules

Total number of electron pairs around central atom	Number of bonding pairs of electrons	Number of lone pairs of electrons	Shape	Bond angle	Examples
2	2	0	Linear	180°	$BeCl_2$, CO_2
3	3	0	Trigonal planar	120°	BF_3
4	4	0	Tetrahedral	109.5°	CH_4, NH_4^+
4	3	1	Pyramidal	107°	NH_3, H_3O^+
4	2	2	Bent	104.5°	H_2O, BrF_2^+
5	5	0	Trigonal bipyramid	90° and 120°	PF_5
5	3	2	T shaped	86°	BrF_3
5	2	3	Linear	180°	XeF_2
6	6	0	Octahedral	90°	SF_6
6	4	2	Square planar	90°	BrF_4^-

■ Redox

Redox is oxidation and reduction occurring simultaneously in the same reaction. There are four different definitions of both oxidation and reduction:

- Oxidation is:
  - loss of electrons
  - gain of oxygen
  - loss of hydrogen
  - increase in oxidation state
- Reduction is:
  - gain of electrons
  - loss of oxygen
  - gain of hydrogen
  - decrease in oxidation state

Oxidation state is the charge on a simple ion or the difference in the number of electrons associated with an element in a compound compared with the atoms of the element.

The **oxidation state** of a particular atom in a compound or ion is a measure of the number of electrons lost, gained or shared.

Working with oxidation states

There are rules for working with oxidation states.

1 The oxidation state of the atoms in an element is 0 (zero). For example:
  - the oxidation state of Na atoms in sodium metal is 0
  - the oxidation state of both Cl atoms in Cl_2 is 0
  - the oxidation state of all eight S atoms in S_8 is 0

2 Oxidation states are always written as positive or negative integers (whole numbers), for example +2, −1, +7.

3 Fractional oxidation states are possible but only as the average of several atoms of the same element in a compound, for example in Fe_3O_4 the oxidation state of iron is $+2\frac{2}{3}$, as two of the irons have an oxidation state of +3 and one has an oxidation state of +2 — the average is $+2\frac{2}{3}$.

4 Oxygen has an oxidation state of −2 in almost all compounds except in peroxides (for example, in hydrogen peroxide (H_2O_2) or sodium peroxide (Na_2O_2) where the oxidation state is −1) and in the compound F_2O where it is +2 (fluorine is more electronegative than oxygen).

5 Hydrogen has an oxidation state of +1 in almost all compounds except metal hydrides, for example NaH and CaH_2 (where it is −1).

6 Group I elements have an oxidation state of +1 in all compounds.

7 Group II elements have an oxidation state of +2 in all compounds.

8 The oxidation state of *simple* ions in a compound is equal to the charge on the ion. For example:
  - in iron(II) chloride, iron has an oxidation state of +2
  - in copper(II) sulfate, copper has an oxidation state of +2
  - in silver(I) nitrate, silver has an oxidation state of +1
  - in sodium chloride, chlorine has an oxidation state of −1
  - in magnesium oxide, oxygen has an oxidation state of −2

9 The total of the oxidation states for the elements in a compound must equal 0 (zero).

10 The total of the oxidation states of the elements in a molecular ion must equal the charge of the ion. For example:
  - the total of the oxidation states of the elements in sulfate, SO_4^{2-}, must be −2
  - the total of the oxidation states of the elements in nitrate, NO_3^- must equal −1

11 The oxidation states of the p and d block elements vary significantly.

12 The maximum oxidation state of a p block element is '+ group number'. For example:
  - the maximum oxidation state of Cl is +7
  - the maximum oxidation state of N is +5

13 The minimum oxidation state of a p block element is 'group number − 8'. For example:
  - the minimum oxidation state of Cl = 7 − 8 = −1
  - the minimum oxidation state of N = 5 − 8 = −3

14 The oxidation states of the d block elements can vary up to +7.

Exam tip

Oxidation number is the same as oxidation state and you may see this in questions.

Calculating oxidation state

The above rules for working with oxidation states are used to calculate oxidation states of different elements in compounds and ions.

Worked example 1

Determine the oxidation state of S in sodium sulfate, Na_2SO_4.

Answer

Na oxidation state = +1   two Na atoms present, so total for Na = +2

S   oxidation state = x (the unknown)

O   oxidation state = −2   four O atoms present, so total for O = −8

Na_2SO_4

$+2 + x − 8 = 0$ (total of oxidation state is zero because it is a compound)

$x = +8 − 2 = +6$

Oxidation state of S in Na_2SO_4 is +6.

Worked example 2

Determine the oxidation state of Mn in potassium permanganate, $KMnO_4$.

Answer

K   oxidation state = +1

Mn oxidation state = x

O   oxidation state = −2   four O atoms present, so total = −8

$KMnO_4$

$+1 + x − 8 = 0$ (total of oxidation state is zero because it is a compound)

$x = +8 − 1 = +7$

Oxidation state of Mn in $KMnO_4$ = +7.

Exam tip

The correct chemical name of potassium permanganate is potassium manganate(VII) where VII represents the +7 oxidation state of manganese in the compound. The dichromate ion is the dichromate(VI) ion.

Worked example 3

Determine the oxidation state of Cr in the dichromate ion, $Cr_2O_7{}^{2-}$.

Answer

Cr oxidation state = x two Cr atoms present, so $2x$

O   oxidation state = −2   seven O atoms present, so total = −14

$Cr_2O_7{}^{2-}$

$2x − 14 = −2$ (total of oxidation state is −2 because it is an ion)

$2x = −2 + 14 = +12$ $x = +6$

Oxidation state of each Cr atom in $Cr_2O_7{}^{2-}$ is +6.

Knowledge check 23

What is the oxidation state of nitrogen in nitrate, $NO_3{}^-$?

Explaining redox

Remember it is often d block and p block elements that are oxidised and reduced, but watch out for elements from other groups oxidised from/reduced to zero oxidation state.

When given a redox equation and asked to explain why it is described as redox, you need to calculate the oxidation states of the elements that are oxidised or reduced (Figure 52).

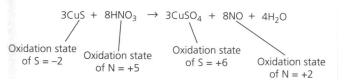

$$3CuS + 8HNO_3 \rightarrow 3CuSO_4 + 8NO + 4H_2O$$

Oxidation state of S = −2
Oxidation state of N = +5
Oxidation state of S = +6
Oxidation state of N = +2

- Sulfur is oxidised from −2 to +6
- Nitrogen is reduced from +5 to +2
- Redox is where oxidation and reduction occur in the same reaction

Figure 52 A redox reaction

Proper names for compounds

In sodium sulfate the oxidation state of the sulfur is +6, so the proper name for sodium sulfate is sodium sulfate(VI). The VI represents the oxidation state of the S in sulfate.

In the dichromate ion, $Cr_2O_7^{2-}$, the oxidation state of both chromium atoms is +6, so the proper name for the dichromate ion is dichromate(VI).

Determining the number of oxygen atoms in an ion or compound

In potassium stannate(IV), K_4SnO_x, the tin has the +4 oxidation state. The number of oxygen atoms x in the compound can be determined.

Worked example

Determine the formula of potassium stannate(IV), K_4SnO_x, where x represents the number of oxygen atoms.

Answer

Each potassium is +1, the tin is +4 and each oxygen is −2, so the total for oxygen is −2x. Since it is a compound, the overall total for the oxidation states is zero (Figure 53).

Solving for x: $-2x = -4 - 4 = -8; x = 4$

Formula of potassium stannate(IV) is K_4SnO_4.

$$K_4SnO_x$$
$$+4 + 4 - 2x = 0$$

Figure 53

Oxidising agent and reducing agent

- An oxidant or **oxidising agent** is a chemical that causes an oxidation in another species.
- Oxidising agents accept electrons (they cause another substance to lose electrons, so oxidising it).
- A reductant or **reducing agent** is a chemical that causes a reduction in another species.
- Reducing agents lose electrons (they cause another substance to gain electrons, so reducing it).

An **oxidising agent** is an electron acceptor.

A **reducing agent** is an electron donor.

Worked example 1

Chlorine reacts with sodium bromide according to the following equation:

$$Cl_2 + 2NaBr \rightarrow 2NaCl + Br_2$$

Br is oxidised from −1 (in NaBr) to 0 (in Br_2). Cl is reduced from 0 (in Cl_2) to −1 (in NaCl).

Chlorine causes the oxidation of bromide ions to bromine. Chlorine is the oxidising agent (or oxidant).

Worked example 2

Hydrogen peroxide reacts with iodide ions.

$$H_2O_2 + 2I^- + 2H^+ \rightarrow 2H_2O + I_2$$

I is oxidised from −1 (in I^-) to 0 (in I_2). O is reduced from −1 (in H_2O_2) to −2 (in H_2O).

Hydrogen peroxide causes the oxidation of iodide ions to iodine. Hydrogen peroxide is the oxidising agent (or oxidant).

Worked example 3

Zinc reacts with ammonium vanadate(v) in acid conditions as follows:

$$3Zn + 2NH_4VO_3 + 6H_2SO_4 \rightarrow 3ZnSO_4 + 2VSO_4 + (NH_4)_2SO_4 + 6H_2O$$

Zn is oxidised from 0 (in Zn) to +2 (in $ZnSO_4$). V is reduced from +5 (NH_4VO_3) to +2 (in VSO_4).

Zinc causes the reduction of vanadium from the +5 oxidation state to +2. Zinc is the reducing agent (or reductant).

Half-equations

A half-equation is an oxidation or reduction equation involving loss of gain of electrons. Examples of simple half-equations are:

$$Mg \rightarrow Mg^{2+} + 2e^-$$

This is an oxidation because 1 mol of magnesium atoms loses 2 mol of electrons to form 1 mol of magnesium ions.

$$Cl_2 + 2e^- \rightarrow 2Cl^-$$

This is a reduction because 1 mol of chlorine molecules gains 2 mol of electrons to form 2 mol of chloride ions.

$$Fe^{2+} \rightarrow Fe^{3+} + e^-$$

This is an oxidation as 1 mol of iron(II) ions loses 1 mol of electrons to form 1 mol of iron(III) ions.

Note: if the equation is oxidation the electrons are on the right-hand side. If the equation is reduction the electrons are on the left-hand side.

More complex half-equations involve calculation of oxidation states and balancing any oxygen atoms gained or lost using H^+ ions and water.

Worked example 1

Manganate(VII), MnO_4^-, can be reduced to manganese(II), Mn^{2+}.

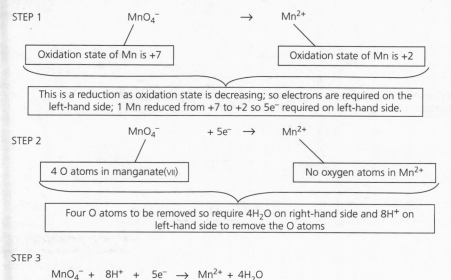

Figure 54 Reduction of manganate(VII) to manganese(II)

Worked example 2

Dichromate(VI), $Cr_2O_7^{2-}$, can be reduced to chromium(III), Cr^{3+}

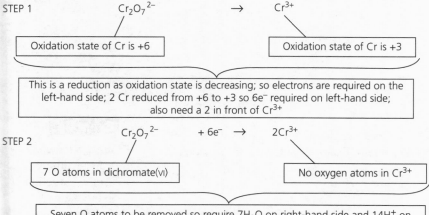

Figure 55 Reduction of dichromate(VI) to chromium(III)

Oxidation of sulfur dioxide, SO_2, to sulfate(vi), SO_4^{2-}.

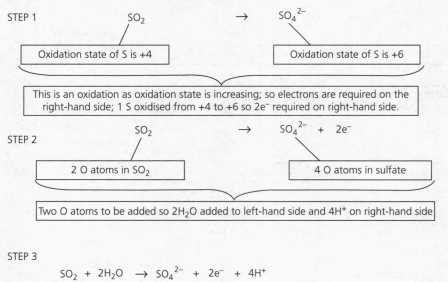

STEP 1 SO_2 $\rightarrow$ SO_4^{2-}

Oxidation state of S is +4

Oxidation state of S is +6

This is an oxidation as oxidation state is increasing; so electrons are required on the right-hand side; 1 S oxidised from +4 to +6 so $2e^-$ required on right-hand side.

STEP 2 SO_2 $\rightarrow$ SO_4^{2-} + $2e^-$

2 O atoms in SO_2

4 O atoms in sulfate

Two O atoms to be added so $2H_2O$ added to left-hand side and $4H^+$ on right-hand side

STEP 3

$$SO_2 + 2H_2O \rightarrow SO_4^{2-} + 2e^- + 4H^+$$

Figure 56 Oxidation of sulfur dioxide to sulfate(vi)

Write a half-equation for the reduction of sulfate(iv) ions to sulfur.

Balancing redox equations

Half-equations include electrons. A half-equation is half a redox reaction and involves the oxidation or reduction of one particular species. Examples of half-equations are:

$Ni \rightarrow Ni^{2+} + 2e^-$

$Cr_2O_7^{2-} + 14H^+ + 6e^- \rightarrow 2Cr^{3+} + 7H_2O$

Features of half-equations:
- They involve electrons.
- Only one species is oxidised or reduced.

Redox ionic equations do not include electrons. A redox ionic equation is the reaction between two ionic species transferring electrons. Examples of redox ionic equations are:

$Mg + 2H^+ \rightarrow Mg^{2+} + H_2$

$Cl_2 + 2I^- \rightarrow 2Cl^- + I_2$

Features of redox ionic equations:
- They do not involve electrons.
- Two species are involved — one oxidised, one reduced.

Often two half-equations are given to you and you are asked to write the ionic equation. This is simply a matter of multiplying the half-equations by a number that gives the same number of electrons in the oxidation half-equation and in the reduction half-equation.

When the equations are added together to make an ionic equation, there will be the same number of electrons on both sides of the ionic equation so they can be cancelled out.

A redox ionic equation is often simply called a redox equation or an ionic equation. However, some ionic equations such as those for precipitation reactions do not involve redox.

There are two ways in which half-equations are presented at AS and A2:

- You may be given one half-equation that is a reduction and the other will be written as an oxidation.
- You may be given two half-equations that are both written as reductions. In this type, one equation needs to be reversed to make it an oxidation before you can add the equations together.

Adding half-equations together

Make sure you have one oxidation equation (electrons on the right) and one reduction (electrons on the left). For example:

$Al \rightarrow Al^{3+} + 3e^-$ oxidation

$F_2 + 2e^- \rightarrow 2F^-$ reduction

Make sure that the numbers of electrons are the same in both the oxidation and reduction half-equations. To do this the oxidation equation needs to be multiplied by 2. The reduction equation needs to be multiplied by 3. This will give both equations six electrons.

$2Al \rightarrow 2Al^{3+} + 6e^-$

$3F_2 + 6e^- \rightarrow 6F^-$

To add them, simply write down all the species from the left-hand side of both half-equations, then put an arrow and finally write down all the species from the right-hand side of both half-equations:

$2Al + 3F_2 + 6e^- \rightarrow 2Al^{3+} + 6e^- + 6F^-$

The next step is to cancel out the electrons on both sides of the ionic equation:

$2Al + 3F_2 \rightarrow 2Al^{3+} + 6F^-$

This is the ionic equation for the reaction between aluminium and fluorine.

Worked example 1

Iron(II) ions are oxidised by acidified potassium manganate(VII). The two half-equations are:

$Fe^{2+} \rightarrow Fe^{3+} + e^-$ oxidation

$MnO_4^- + 8H^+ + 5e^- \rightarrow Mn^{2+} + 4H_2O$ reduction

There are five electrons on the left-hand side of the second equation and only one electron on the right-hand side of the first equation.

To write a complete ionic equation, the first equation must be multiplied by 5 and then the two equations are simply added together:

$5Fe^{2+} \rightarrow 5Fe^{3+} + 5e^-$

$MnO_4^- + 8H^+ + 5e^- \rightarrow Mn^{2+} + 4H_2O$

$MnO_4^- + 8H^+ + 5e^- + 5Fe^{2+} \rightarrow Mn^{2+} + 4H_2O + 5Fe^{3+} + 5e^-$ →

Exam tip

Always check an ionic equation for charges on the left-hand side and on the right-hand side — the total charges should be the same on both sides.

The $5e^-$ on each side can be cancelled so that the overall equation reads:

$$MnO_4^- + 8H^+ + 5Fe^{2+} \rightarrow Mn^{2+} + 4H_2O + 5Fe^{3+}$$

Sometimes the half-equations are given as two reductions, so one equation must be reversed to enable the electrons to be eliminated. The reaction will indicate which two species are reacting. The following could be the way in which the above example was presented:

$$Fe^{3+} + e^- \rightarrow Fe^{2+} \quad \text{reduction}$$

$$MnO_4^- + 8H^+ + 5e^- \rightarrow Mn^{2+} + 4H_2O \qquad \text{reduction}$$

You are asked to write an ionic equation for the reaction between iron(II) ions, Fe^{2+}, and manganate(VII) ions, MnO_4^-. The reaction requires the first half-equation to be reversed to:

$$Fe^{2+} \rightarrow Fe^{3+} + e^-$$

and then multiplied by 5 as before.

The final equation is the same when the electrons have been eliminated:

$$MnO_4^- + 8H^+ + 5Fe^{2+} \rightarrow Mn^{2+} + 4H_2O + 5Fe^{3+}$$

Worked example 2

Write an ionic equation for the reaction of nitrate(III) ions, NO_2^-, and dichromate(VI) ions, $Cr_2O_7^{2-}$, using the following half-equations:

$$NO_2^- + H_2O \rightarrow NO_3^- + 2H^+ + 2e^-$$

$$Cr_2O_7^{2-} + 14H^+ + 6e^- \rightarrow 2Cr^{3+} + 7H_2O$$

The equations given are an oxidation (first equation) and a reduction (second equation) so they can be combined directly once the electrons have been balanced.

$$3NO_2^- + 3H_2O \rightarrow 3NO_3^- + 6H^+ + 6e^-$$

$$Cr_2O_7^{2-} + 14H^+ + 6e^- \rightarrow 2Cr^{3+} + 7H_2O$$

$$Cr_2O_7^{2-} + 14H^+ + 3NO_2^- + 3H_2O + 6e^- \rightarrow 2Cr^{3+} + 7H_2O + 3NO_3^- + 6H^+ + 6e^-$$

The electrons, water and H^+ need to be cancelled down on both sides:

$$Cr_2O_7^{2-} + 8H^+ + 3NO_2^- \rightarrow 2Cr^{3+} + 4H_2O + 3NO_3^-$$

Disproportionation

A **disproportionation** reaction is one in which the same element is oxidised and reduced. An example is shown in Figure 57.

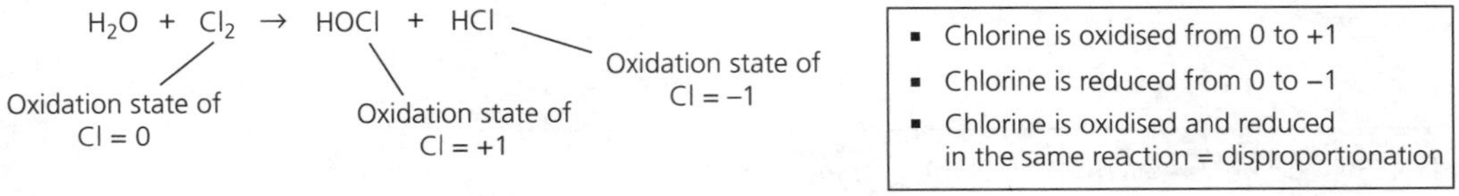

Figure 57 Disproportionation

> **Knowledge check 25**
>
> What is meant by the term disproportionation?
>
> **Disproportionation** is oxidation and reduction of the same element in the same reaction.

Summary

- The oxidation state (or oxidation number) of a particular element in a compound or ion is a measure of the number of electrons lost or gained or shared.
- Hydrogen has an oxidation state of +1 in almost all compounds and ions except hydrides.
- Oxygen has an oxidation state of −2 in almost all compounds and ions except peroxides and F_2O.
- In a compound, group I elements have an oxidation number of +1; group II elements +2 and Al +3.
- Transition metals in compounds and ions vary in their oxidation state.
- Oxidation states should always be stated with a + or − sign.
- The total oxidation state in a compound equals zero.
- The total oxidation state in a molecular ion equals the charge on the ion.
- Oxidising agents gain electrons and cause another species to lose electrons and become oxidised.
- Reducing agents lose electrons and cause another species to gain electrons and become reduced.
- Half-equations represent oxidation and reduction; reduction is gain of electrons and oxidation is loss of electrons.
- When combining half-equations the number of electrons are made to cancel out and any H^+ ions and H_2O on both sides of the equation can cancel down.

Halogens

Physical properties

Table 11 gives some of the physical properties of the first four halogens.

Table 11

Halogen	Fluorine	Chlorine	Bromine	Iodine
Colour and physical state at room temperature	Yellow gas	Yellow-green/ green-yellow/ green gas	Red-brown liquid (red-brown vapour when heated)	Grey-black solid (violet/purple vapour when heated)
Melting point/°C	−220	−101	−7	114
Boiling point/°C	−188	−34	59	184
Atomic radius/nm	0.057	0.01	0.115	0.140
First ionisation energy/ $kJ\,mol^{-1}$	1681	1251	1140	1008
Electronegativity	4.0	3.0	2.8	2.5

Exam tip

You need to be able to explain the physical properties of the halogens using your knowledge of previous sections. For example, the increase in melting point, boiling point and the change in physical state can be explained in terms of van der Waals forces. The increasing atomic radius, decreasing first ionisation energy and electronegativity can be explained in terms of the outer electrons being further from the nucleus.

Solubility of the halogens

The halogens are non-polar simple covalent molecules. The solubility of the halogens in water (a polar solvent) decreases down the group, until iodine, which is virtually insoluble in water. A solution of chlorine in water is called 'chlorine water' and a solution of bromine in water is called 'bromine water'.

All of the halogens dissolve in non-polar (non-aqueous) solvents such as hexane (Table 12).

Table 12 Solubility of the halogens in water and in hexane

Halogen	Solubility in water	Solubility in hexane
Chlorine	Soluble, forming a green/colourless solution	Soluble, forming a colourless solution
Bromine	Soluble, forming a yellow/orange/brown solution	Soluble, forming a red solution
Iodine	Virtually insoluble, but any solution formed is yellow/brown (iodine is soluble in a solution containing iodide ions to form a brown solution; it is soluble in other polar solvents, such as ethanol, to form a yellow/brown solution)	Soluble, forming a purple solution

Halogens reacting with water

$$Cl_2 + H_2O \rightarrow HOCl + HCl$$

Chlorine is oxidised from an oxidation state of 0 in Cl_2 to +1 in HOCl and is also reduced from an oxidation state of 0 in Cl_2 to −1 in HCl. This is a disproportionation reaction (see p. 58) where one species is oxidised and reduced in the same reaction. The resulting solution contains hydrogen ions, chloride ions and chlorate(I) ions.

> **Exam tip**
>
> When describing a redox or disproportionation reaction, remember to include the change in oxidation state and the substances in which these oxidation states are found. Remember that an increase in oxidation state is oxidation and a decrease in oxidation state is reduction. Redox is oxidation and reduction occurring in the same reaction. Disproportionation is oxidation and reduction of the same element in the same reaction.

Halogens reacting with alkali

Reaction with cold dilute sodium hydroxide solution

Chlorine reacts with cold dilute sodium hydroxide solution according to the equation:

$$2NaOH + Cl_2 \rightarrow NaCl + NaClO + H_2O$$

The general ionic equation for this reaction is:

$$2OH^- + Cl_2 \rightarrow Cl^- + ClO^- + H_2O$$

Knowledge check 26

What is the chemical name for NaClO?

NaClO is called sodium hypochlorite or sodium chlorate(I) as it contains the chlorate(I) ion, ClO^-. The oxidation number of chlorine in chlorate(I) is +1.

Reaction with hot concentrated sodium hydroxide solution

Chlorine reacts with hot concentrated sodium hydroxide solution according to the equation:

$$3Cl_2 + 6NaOH \rightarrow 5NaCl + NaClO_3 + 3H_2O$$

The general ionic equation for this reaction is:

$$6OH^- + 3Cl_2 \rightarrow 5Cl^- + ClO_3^- + 3H_2O$$

$NaClO_3$ is sodium chlorate(v) or simply 'sodium chlorate' since it contains the chlorate(v) ion, ClO_3^-. The oxidation number of chlorine in chlorate(v) is +5.

Bromine reacts with sodium hydroxide in the same way as chlorine.

Iodine does not react with cold dilute alkali, but will react with hot concentrated alkali to form an iodate(v) compound.

Again in these reactions chlorine undergoes disproportionation since it is both oxidised and reduced in the same reaction: Cl_2 (oxidation number = 0); Cl in Cl^- (oxidation number = −1); Cl in ClO_3^- (oxidation number = +5).

In all of the above reactions, the yellow-green gas dissolves to form a colourless solution.

Oxidising ability of the halogens

■ The oxidising power of the halogens decreases as atomic number increases from fluorine to chlorine to bromine to iodine.
■ Oxidising agents (also called oxidants) cause an oxidation to occur so they cause another atom, molecule or ion to lose electrons. This means that an oxidising agent gains electrons and becomes reduced.
■ Oxidising agents readily accept electrons.
■ The ease with which an oxidising agent gains electrons determines how effective it is as an oxidising agent.
■ The halogen atoms gain electrons in their outer energy level to complete the energy level.
■ The electron gained by a fluorine atom completes an energy level closer to the nucleus than the electron that completes the outer energy level in chlorine.
■ The electron that fluorine gains has a stronger attraction to the nucleus because it is closer to the nucleus than for a chlorine atom.
■ The electron gained by a fluorine atom is not subjected to as much shielding by inner electrons because there are fewer electrons between the electron gained and the nucleus.

> **Exam tip**
>
> The reactions of fluorine are not examined experimentally because fluorine is too dangerous to be used in the laboratory. The reactions of fluorine would follow the pattern for the other halogens.

Displacement reactions

From our knowledge of the oxidising powers of the halogens we would expect fluorine to displace all other halides from a solution of a halide compound. We would also expect chlorine to displace bromide and iodide from solutions of halide compounds, and bromine to displace iodide in solution.

Worked example 1

Write the simplest ionic equation for the reaction of chlorine with sodium bromide solution. Write half-equations for the oxidation and reduction reactions that occur.

Answer

The equation for this reaction is:

$$Cl_2 + 2NaBr \rightarrow 2NaCl + Br_2$$

The Na^+ ion is a spectator ion so it is not included in the ionic equation:

$$Cl_2 + 2Br^- \rightarrow 2Cl^- + Br_2$$

The half-equation for the conversion of chlorine molecules to chloride ions is:

$$Cl_2 + 2e^- \rightarrow 2Cl^-$$

This is a reduction half-equation because chlorine gains electrons.

The half-equation for the conversion of bromide ions to bromine molecules is:

$$2Br^- \rightarrow Br_2 + 2e^-$$

this is an oxidation half-equation because the bromide ions lose electrons.

Chlorine reacts with sodium iodide solution to form sodium chloride and iodine:

$$2NaI + Cl_2 \rightarrow 2NaCl + I_2$$

The solution changes colour from colourless to yellow/brown.

Worked example 2

Bromine reacts with a solution containing iodide ions such as potassium iodide solution. This reaction is not used as extensively as the other two above because the colour change in the solution is not as clearcut — it simply darkens from orange to brown:

$$Br_2 + 2I^- \rightarrow 2Br^- + I_2$$

Half-equations:

$$Br_2 + 2e^- \rightarrow 2Br^- \quad \text{reduction}$$

$$2I^- \rightarrow I_2 + 2e^- \quad \text{oxidation}$$

Exam tip

You may be asked to write half-equations for the conversion of molecules to ions or of ions to molecules. You may also have to identify these reactions as oxidation or reduction processes.

Exam tip

In this reaction the colourless solution changes to orange because bromine is produced in the solution and bromine water is yellow/orange/brown.

Knowledge check 27

Which of chlorine, bromine or iodine is the strongest oxidising agent?

Reducing ability of the halides

- The reducing power of the halide ions increases as atomic number increases from fluoride to chloride to bromide to iodide.
- Reducing agents (also called reductants) cause a reduction to occur so they cause another atom, molecule or ion to gain electrons. This means that a reducing agent loses electrons and becomes oxidised.
- Reducing agents readily donate electrons.
- The ease with which a reducing agent loses electrons determines how effective it is as a reducing agent.
- Halide ions lose electrons from their outer energy level.
- The electron that is lost by an iodide ion comes from an energy level further from the nucleus than the electron that would be lost by a bromide ion.
- The electron that an iodide ion loses has a weaker attraction to the nucleus because it is further from the nucleus compared with the other halide ions.
- The electron lost by an iodide ion is subjected to greater shielding by inner electrons than other halide ions because there are more electrons between the electron being lost and the nucleus.
- Fluoride and chloride ions have little reducing ability because they are not as able to donate electrons as bromide and iodide ions.

Knowledge check 28

Which of chloride, bromide or iodide is the strongest reducing agent?

Halides and concentrated sulfuric acid

Solid halide compounds react with concentrated sulfuric acid. The equations, observations and names of the products of these reactions are common questions.

> **Exam tip**
>
> There are seven equations but the first equation in each is the same, just with a different halide ion and hydrogen halide. The second equation for bromide and iodide is also the same, again just change the halide and hydrogen halide. This means you only have to remember four equations rather than seven.

Fluoride with concentrated sulfuric acid

Equation: $NaF + H_2SO_4 \rightarrow NaHSO_4 + HF$

This is not a redox reaction; HF has no reducing ability. HSO_4^- is the hydrogensulfate ion.

Observations: misty fumes (HF); heat released; gas produced; solid disappears; pungent smell (HF)

Names of products: sodium hydrogensulfate and hydrogen fluoride

> **Exam tip**
>
> The reactions of solid fluorides should not be carried out in the laboratory because HF is highly dangerous, forming corrosive and penetrating hydrofluoric acid in contact with moisture.

Chloride with concentrated sulfuric acid

Equation: $NaCl + H_2SO_4 \rightarrow NaHSO_4 + HCl$

This is not a redox reaction; HCl has no reducing ability.

Observations: misty fumes (HCl); heat released; gas produced; solid disappears; pungent smell (HCl)

Names of products: sodium hydrogensulfate and hydrogen chloride

The reaction with a solid fluoride is similar to that of the chloride, releasing misty fumes of hydrogen fluoride, but it is too dangerous to be carried out in a school laboratory.

Bromide with concentrated sulfuric acid

Equations: $NaBr + H_2SO_4 \rightarrow NaHSO_4 + HBr$

$2HBr + H_2SO_4 \rightarrow Br_2 + SO_2 + 2H_2O$

HBr has some reducing ability, so H_2SO_4 is reduced to SO_2 and bromide is oxidised to Br_2.

Observations: misty fumes (HBr); heat released; gas produced; solid disappears; red-brown vapour (Br_2); pungent smell (HBr/SO_2/Br_2)

Names of products: sodium hydrogensulfate, hydrogen bromide, bromine, sulfur dioxide and water

Iodide with concentrated sulfuric acid

Equations: $NaI + H_2SO_4 \rightarrow NaHSO_4 + HI$

$2HI + H_2SO_4 \rightarrow SO_2 + I_2 + 2H_2O$

$6HI + H_2SO_4 \rightarrow S + 3I_2 + 4H_2O$

$8HI + H_2SO_4 \rightarrow H_2S + 4I_2 + 4H_2O$

HI has good reducing ability, so H_2SO_4 is reduced to SO_2, S and H_2S while iodide is oxidised to I_2.

Observations: misty fumes (HI); heat released; gas produced; solid disappears; purple vapour and grey-black solid (I_2); pungent smell (HI/SO_2/I_2); rotten eggs smell (H_2S); yellow solid (S)

Names of products: sodium hydrogensulfate; hydrogen iodide; iodine; sulfur dioxide; sulfur; hydrogen sulfide; water

The reactions with concentrated sulfuric can be used as an alternative test for a halide ion (see page 77 for testing for halide ions using silver nitrate solution):

- chloride ions give hydrogen chloride gas with concentrated sulfuric acid, which can be tested using a glass rod dipped in concentrated ammonia solution — white smoke given off
- bromide ions give bromine vapour with concentrated sulfuric acid, which can be seen as red-brown fumes
- iodide ions give iodine vapour with concentrated sulfuric acid, which can be seen as purple fumes and/or a grey-black solid is formed

Knowledge check 29

Write two equations for the reaction of concentrated sulfuric acid with potassium bromide.

Halides and concentrated phosphoric acid

The hydrogen halides cannot reduce phosphoric acid. Solid metal halides react with concentrated phosphoric acid to form the hydrogen halide and the dihydrogenphosphate salt. The dihydrogenphosphate ion is $H_2PO_4^-$.

Misty fumes of the hydrogen halide are observed.

Sodium fluoride, sodium chloride, potassium bromide and potassium iodide react as follows:

$$NaF + H_3PO_4 \rightarrow NaH_2PO_4 + HF$$

$$NaCl + H_3PO_4 \rightarrow NaH_2PO_4 + HCl$$

$$KBr + H_3PO_4 \rightarrow KH_2PO_4 + HBr$$

$$KI + H_3PO_4 \rightarrow KH_2PO_4 + HI$$

There are no further reactions with phosphoric acid.

Chlorine and ozone in drinking water

Chlorine or ozone may be added to drinking water at source to kill microorganisms.

Ozone, O_3, is an allotrope of molecular oxygen, O_2. It can be generated by passing a high voltage electric discharge through oxygen, or on a smaller scale using ultraviolet light:

$$3O_2 \rightarrow 2O_3$$

Chlorine reacts with water to form a mixture of hypochlorous acid (HOCl) and hydrochloric acid (HCl):

$$Cl_2 + H_2O \rightarrow HOCl + HCl$$

Both ozone and HOCl kill microorganisms by oxidising the microorganisms.

Table 13 details some of the advantages and disadvantages of using chlorine and ozone in the treatment of drinking water.

Table 13

	Advantages	Disadvantages
Chlorine	■ Cheaper than ozone ■ Provides residual protection as it is still present in water when it reaches the consumer ■ More soluble in water than ozone	■ Cannot kill some microorganisms ■ Leaves chemicals in the water ■ Unpleasant taste ■ Toxic to humans, so very low quantities used
Ozone	■ Kills more different types of microorganisms than chlorine ■ Breakdown product is oxygen ■ No residual chemicals in the water ■ Removes metal and organic particles from water ■ No unpleasant taste	■ More expensive than chlorine ■ Does not provide residual protection against microorganisms ■ Less soluble in water than chlorine so requires special mixing techniques

Summary

- Group VII comprises the halogens, which are coloured, non-metallic, reactive elements.
- The halogens react with some other halide ions in solution.
- The oxidising power of the halogens decreases down the group; the reducing ability of the halide ions and the hydrogen halides increases down the group.
- Solid halide compounds (chlorides, bromides and iodides) react with concentrated sulfuric acid.
- Hydrogen bromide and hydrogen iodide can reduce concentrated sulfuric acid (HF and HCl cannot).
- None of the hydrogen halides can reduce concentrated phosphoric acid.
- Chlorine (Cl_2) and ozone (O_3) can be used in water treatment to kill microorganisms.

■Acid–base titrations

A titration is a method of volumetric analysis. One solution is placed in a burette and the other is placed in a conical flask. An indicator is added to the solution in the conical flask. The solution in the burette is added to the solution in the conical flask. The indicator will show the end-point of the titration (when the indicator changes colour). This is the point when the reaction is complete.

Safety, accuracy and reliability

Throughout the process of a titration, safety, accuracy and reliability are paramount and are often examined.

The following features are important for *safety*:
- the use of a pipette filler
- the use of gloves (where appropriate)

The following features of a titration are important for *accuracy*:
- rinse the apparatus with the appropriate solution
- add the solution dropwise near the end-point
- swirl the flask and wash down the sides of the flask with distilled/deionised water
- read the burette at the bottom of the meniscus

The following features of a titration are important for **reliability**:
- repeat the titration two or three times
- concordant readings should be obtained (within $0.1\ cm^3$ of each other)

Apparatus and practical techniques

The main pieces of apparatus used in a titration are a burette, a pipette with safety filler, a volumetric flask and several conical flasks.

Preparing a burette for use
- Rinse the burette with deionised water.
- Ensure the water flows through the jet.
- Discard the water.

Exam tip

When using concentrated solutions of acid or alkali where you are told they are corrosive, gloves should be worn.

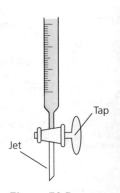

Tap

Jet

Figure 58 Burette

- Rinse the burette with the solution you will be filling it with.
- Ensure the solution flows through the jet.
- Discard the solution.
- Charge (fill) the burette with the solution you will be using in it, ensuring that the jet is filled.

Using a burette

When using a burette the volume of solution it contains is read at the bottom of the meniscus as shown in Figure 59. If you are right-handed, the tap of the burette is operated with the left hand to allow the right hand to be used to swirl to mix the contents of the conical flask.

Preparing a pipette for use in a titration

- Use a pipette filler.
- Rinse the pipette using deionised water.
- Discard the water.
- Rinse the pipette with the solution you will be filling it with.
- Discard this solution.

Using a pipette

A pipette accurately measures an exact volume of a solution and should be used in the following way:

- A pipette filler is attached to the top of a pipette.
- The pipette is placed in the solution and suction applied to draw the solution up.
- The solution is drawn up above the line on the pipette.
- The solution is released until the bottom of the meniscus sits on the line.
- The solution in the pipette is released into a conical flask.

Pipettes measure out exactly $25.0\,cm^3$ or $10.0\,cm^3$.

(Remember that $1\,cm^3 = 1\,ml$ but chemists prefer cm^3 as their volume unit.)

An exact volume of a solution is vital in volumetric work. Taking *exactly* $25.0\,cm^3$ of a known concentration of a solution means that we know exactly how many moles of the dissolved substance are present in the conical flask.

Conical flasks

Conical flasks are used in titrations as they can be swirled easily to mix the reactants. The sloped sides prevent any of the solution spitting out when it is being added. The conical flask should be rinsed out with deionised water before use.

The conical flask does not have to be completely dry before use because the exact volume of solution added contains an exact number of moles of solute. Extra deionised water does not add to the number of moles of solute.

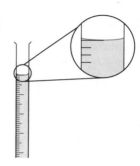

Figure 59 Reading the meniscus

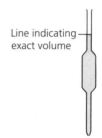

Line indicating exact volume

Figure 60 Pipette

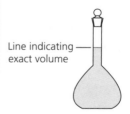

Line indicating exact volume

Figure 61 Volumetric flask

Volumetric flasks

Volumetric flasks are used when diluting one of the solutions before the titration is carried out. They can also be used when preparing a solution of a solid.

Carrying out a dilution of a solution

- Pipette $25.0\,cm^3$ of the original solution into a clean volumetric flask.
- Add deionised water to the flask until the water is just below the line.
- Using a disposable pipette add deionised water very slowly until the bottom of the meniscus is on the line.
- Stopper the flask and invert to mix thoroughly.

Dilution factor

The dilution factor is the amount the original solution is diluted by. It is calculated by dividing the new total volume by the volume of original solution put into the mixture. For example:

- If a $25.0\,cm^3$ sample of a solution is made up to a total volume of $250.0\,cm^3$ using deionised water then the dilution factor is 10.
- If a $10.0\,cm^3$ sample of a solution is made up to a total volume of $250.0\,cm^3$ using deionised water then the dilution factor is 25.

Preparing a solution from a mass of solid/volume of a liquid

When preparing a solution from a solid or a liquid it is important not to lose any of the solid/liquid or the prepared solution before it is placed in the volumetric flask.

- Weigh out an accurate mass of a solid in a weighing boat, or measure out an accurate volume of liquid, and dissolve in a suitable volume of deionised water in a beaker — stirring with a glass rod — then rinse the weighing boat/measuring cylinder/burette/graduated pipette into the beaker with deionised water.
- Dissolve the solid/liquid in a small volume of deionised water; once dissolved, hold the glass rod above the beaker and rinse it with deionised water before removing it.
- Place a glass funnel into the top of a clean volumetric flask and pour the prepared solution down a glass rod into the funnel.
- Rinse the glass rod with deionised water into the funnel.
- Rinse the funnel with deionised water.
- Remove the funnel and add deionised water to the volumetric flask until the water is just below the line.
- Using a disposable pipette add deionised water very slowly until the bottom of the meniscus is on the line.
- Stopper the flask and invert to mix thoroughly.

Knowledge check 30

What is the dilution factor if $25.0\,cm^3$ is diluted to $1\,dm^3$?

Exam tip

You may have to calculate the volume of liquid from its mass using density.

Exam tip

The volume of deionised water must be appropriate to the final volume of solution required. If $250\,cm^3$ of solution is required then a volume of $100\,cm^3$ of deionised water is used so that the rinsings can be accommodated. Adjust the volume between $50\,cm^3$ and $100\,cm^3$ depending on the final volume required.

Carrying out a titration

The major points in carrying out a titration are shown in Figure 62.

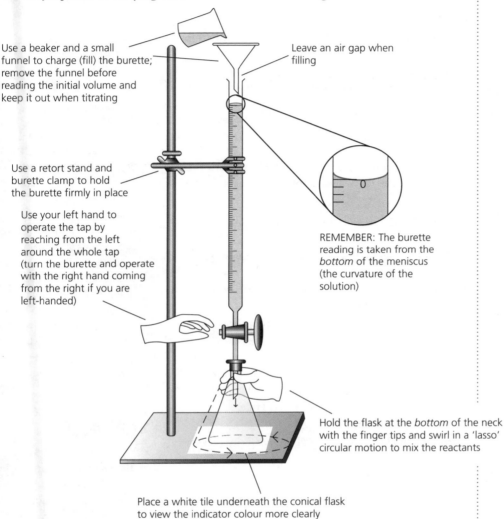

Use a beaker and a small funnel to charge (fill) the burette; remove the funnel before reading the initial volume and keep it out when titrating

Leave an air gap when filling

Use a retort stand and burette clamp to hold the burette firmly in place

Use your left hand to operate the tap by reaching from the left around the whole tap (turn the burette and operate with the right hand coming from the right if you are left-handed)

REMEMBER: The burette reading is taken from the *bottom* of the meniscus (the curvature of the solution)

Hold the flask at the *bottom* of the neck with the finger tips and swirl in a 'lasso' circular motion to mix the reactants

Place a white tile underneath the conical flask to view the indicator colour more clearly

Figure 62 Carrying out a titration

In one volumetric analysis, a minimum of three titrations should be carried out:

- The first titration should be rough and should be an overshoot, but no more than $1\,cm^3$ greater than the accurate titrations.
- The subsequent titrations should be accurate (within $0.10\,cm^3$ of each other) with drop-wise addition as the end-point is reached.

Standard solutions are used in volumetric analysis.

A burette has a total graduated volume of $50.0\,cm^3$. You can perform two titrations using a burette if the titres are well below $25.0\,cm^3$. However, if the rough titration is close to or above $25.0\,cm^3$, it is essential that you refill the burette before starting the first accurate titration.

A **standard solution** is a solution for which the concentration is accurately known.

The funnel is removed to ensure that any drops of solution remaining in the funnel do not interfere with the titration results.

Recording titration results

Table 14 is a typical table used to record titration results in practical examinations.

Table 14

	Rough titration	Accurate titration 1	Accurate titration 2
Initial burette reading/cm^3			
Final burette reading/cm^3			
Titre/cm^3			

The titre is the volume delivered from the burette into the conical flask until the indicator changes colour (end-point). 'Volume delivered' is sometimes used in the table instead of 'titre'.

The following points about recording titration results in a practical examination are very important:

- You should put units with the headings and never put units in the main body of the table.
- All values should be recorded to 1 decimal place — i.e. 0 cm^3 is written as 0.0 cm^3; 25 cm^3 is written as 25.0 cm^3.
- A rough titration titre should be greater than the accurate titration titres but not more than 1 cm^3 greater.
- Concordant titration titre values should be within 0.1 cm^3 of each other. Titrations should be carried out until two values within 0.1 cm^3 of each other are obtained.
- The average titre must be stated with units to a minimum of one decimal place. Two decimal places are acceptable, e.g. 23.95 cm^3 if necessary. However, if the average titre is 24.0 cm^3, do not write 24.00 cm^3.
- Your results are marked according to how closely grouped the accurate titres are.

Calculating the average titre

When calculating the average titre, ignore the rough titration and any result that is clearly not within 0.1 cm^3 of the other accurate titration values.

Write the average titre below the table and include the units. The average titre can be written to two decimal places. The titres must only be to one decimal place. Table 15 shows an example which could be presented as a set of results.

Table 15 Sample titration results table

	Titration 1	Titration 2	Titration 3	Titration 4
Initial burette reading/cm^3	0.0	0.0	0.4	0.4
Final burette reading/cm^3	24.5	24.0	24.3	24.0
Titre/cm^3	24.5	24.0	23.9	23.6

Average titre = 23.95 cm^3

The average titre is calculated from concordant titres. The only two concordant titres here are titrations 2 and 3. You may be asked to identify the titrations that should be used to calculate the average titre. The titres for titrations 1 and 4 are outliers as they are not concordant results. Titration 1 may have been a rough titre.

Units

- Units of volume are in cm^3 ($1\,cm^3 = 1\,ml$).
- Concentration units are $mol\,dm^{-3}$. This can also be written as ' mol/dm^3'.
- Remember that $1\,dm^3$ is the same as 1 litre.
- However, M (molar) may also be used and is the same as $mol\,dm^{-3}$.
- A $1\,M$ solution is $1\,mol\,dm^{-3}$. The molarity of a solution has units M but concentration is usually quoted as $mol\,dm^{-3}$ (it can also be given in $g\,dm^{-3}$).
- Concentration in $g\,dm^{-3}$ (grams per dm^3) can be calculated by multiplying the concentration (or **molarity**) by the RFM of the solute.

Concentration is the number of moles or mass present in a stated volume. **Molarity** is the concentration in $mol\,dm^{-3}$ expressed using M (molar).

Uncertainty in apparatus and measurements

The percentage uncertainty or error in apparatus or measurements can be calculated.

A volume of $25\,cm^3$ measured with a measuring cylinder has an uncertainty of $\pm0.5\,cm^3$. The percentage uncertainty for this measurement is:

$$\frac{0.5}{25} \times 100 = 2\%$$

A volume of $25\,cm^3$ measured using a class A pipette has an uncertainty of $\pm0.03\,cm^3$. The percentage uncertainty is:

$$\frac{0.03}{25} \times 100 = 0.12\%$$

A mass of $0.120\,g$ measured using a balance with an uncertainty of $\pm0.001\,g$ has a percentage uncertainty of:

$$\frac{0.001}{0.120} \times 100 = 0.833\%$$

A $50\,cm^3$ class B burette has an uncertainty of $0.05\,cm^3$. The burette has a percentage uncertainty of:

$$\frac{0.05}{50} \times 100 = 0.1\%$$

When two measurements are used to calculate a value then twice the uncertainty is appropriate. A titre (like a temperature change) depends on two values, so the total uncertainty of a titre of $22.5\,cm^3$ is $0.05 \times 2 = 0.10\,cm^3$. This gives a percentage uncertainty for the titre of:

$$\frac{0.10}{22.5} \times 100 = 0.44\%$$

Common acids and bases

Weak and strong acids and bases are used in titrations. A **strong acid** is fully dissociated in solution. A **weak acid** is partially dissociated in solution. Common examples include:

- strong acids: HCl, H_2SO_4, HNO_3
- weak acids: CH_3COOH, organic acids
- strong bases: $NaOH$, KOH
- weak bases: NH_3, Na_2CO_3

Common equations used in titrations:

$NaOH + HCl \rightarrow NaCl + H_2O$ 1:1 ratio for $NaOH$:HCl

$2NaOH + H_2SO_4 \rightarrow Na_2SO_4 + 2H_2O$ 2:1 ratio for $NaOH$:H_2SO_4

$NaOH + CH_3COOH \rightarrow CH_3COONa + H_2O$ 1:1 ratio for $NaOH$:CH_3COOH

Each of the above can be rewritten with KOH and the same ratios apply.

$Na_2CO_3 + 2HCl \rightarrow 2NaCl + CO_2 + H_2O$ 1:2 ratio for Na_2CO_3:HCl

$Na_2CO_3 + H_2SO_4 \rightarrow Na_2SO_4 + CO_2 + H_2O$ 1:1 ratio for Na_2CO_3:H_2SO_4

The ratio between the solutes is important because it gives the ratio between the number of moles of the substance added from the burette and the number of moles of the other substance in the conical flask.

Indicators

The two main indicators used for acid–base titrations are phenolphthalein and methyl orange (Table 16).

Table 16 Features of phenolphthalein and methyl orange

Indicator	Phenolphthalein	Methyl orange
Colour in acidic solutions	Colourless	Red
Colour in neutral solutions	Colourless	Orange
Colour in alkaline solutions	Pink	Yellow
Titrations suitable for:	Strong acid–strong base Weak acid–strong base	Strong acid–strong base Strong acid–weak base

Exam tip

The choice of indicator is based on the types of acid and base in the titration. It is important to be able to choose the correct indicator for a particular titration. If ethanoic acid (a weak acid) and sodium hydroxide solution (a strong alkali) are being used then phenolphthalein is the indicator of choice. Both methyl orange and phenolphthalein can be used for strong acid–strong base titrations.

A **strong acid** or **strong base** fully dissociates in solution.

A **weak acid** or **weak base** partially dissociates in solution.

Knowledge check 31

State two common strong acids.

The colour change of the indicator at the end-point is frequently asked (Table 17).

Table 17 Colour changes of phenolphthalein and methyl orange

Titration	Phenolphthalein	Methyl orange
Acid in conical flask, alkali in burette	Colourless to pink	Red to yellow
Alkali in conical flask, acid in burette	Pink to colourless	Yellow to red

Typical method of acid–base titration

- The solution of known concentration is usually placed in the burette.
- Using a pipette filler, rinse a pipette with deionised water and with the solution you are going to pipette into the conical flask. Pipette 25.0 cm³ of this solution into three different conical flasks.
- Add 3–5 drops of a suitable indicator to each conical flask.
- Note the colour of the indicator in this solution.
- Rinse a burette with deionised water and with the solution you are using to fill it. Fill the burette with this solution. Titrate until the indicator just changes colour, adding dropwise near the end-point.
- Repeat for accuracy and calculate the average titre from concordant titre values.

Volumetric calculations

Calculations involving solutions are a little more complex because you are dealing with a solution volume and a concentration of the solution to determine the number of moles of solute present. The expression used is:

$$\text{number of moles} = \frac{\text{solution volume (cm}^3) \times \text{concentration (mol dm}^{-3})}{1000}$$

Typical acid–base calculation question

A solution of oven cleaner contains sodium hydroxide. 25.0 cm³ of this solution were pipetted into a 250 cm³ volumetric flask and the solution made up to the mark with deionised water. 25.0 cm³ of this solution were placed in a conical flask and titrated against 0.02 mol dm⁻³ hydrochloric acid using phenolphthalein indicator. The average titre was determined to be 17.2 cm³.

Standard types of part questions asked

a   Calculate the number of moles of hydrochloric acid added from the burette.

b   Write a balanced symbol equation for the reaction between sodium hydroxide and hydrochloric acid.

c   Calculate the number of moles of sodium hydroxide present in 25.0 cm³ of the diluted solution.

From this point the calculation can go one of two ways:

Method 1	Method 2
d Calculate the concentration of the diluted sodium hydroxide solution in $mol\,dm^{-3}$.	d Calculate the number of moles of sodium hydroxide present in $250.0\,cm^3$ of the diluted solution.
e Calculate the concentration of the undiluted sodium hydroxide solution in $mol\,dm^{-3}$.	e Calculate the number of moles of sodium hydroxide present in $25.0\,cm^3$ of the original solution.
f Calculate the concentration of the undiluted sodium hydroxide solution in $g\,dm^{-3}$.	f Calculate the concentration of the undiluted sodium hydroxide solution in $mol\,dm^{-3}$.
	g Calculate the concentration of the undiluted sodium hydroxide solution in $g\,dm^{-3}$.

Answers

a $\quad \text{moles} = \dfrac{\text{solution volume} \times \text{concentration}}{1000} = \dfrac{17.2 \times 0.02}{1000} = 3.44 \times 10^{-4}\,mol$

b $\quad NaOH + HCl \rightarrow NaCl + H_2O$

c $\quad$ 1:1 ratio of NaOH:HCl, so moles of NaOH in $25.0\,cm^3 = 3.44 \times 10^{-4}$

Method 1	Method 2
d concentration = number of moles in $25.0\,cm^3 \times 40$ $= 3.44 \times 10^{-4} \times 40 = 0.01376\,mol\,dm^{-3}$	d Number of moles in $250\,cm^3$ = number of moles in $25\,cm^3 \times 10 = 3.44 \times 10^{-4} \times 10 = 3.44 \times 10^{-3}\,mol$
e Dilution factor is ×10 (25 into $250\,cm^3$) so concentration = $0.01376 \times 10 = 0.1376\,mol\,dm^{-3}$	e This is the same as the number of moles present in $250\,cm^3$ of the diluted solution = $3.44 \times 10^{-3}\,mol$
f Concentration in $g\,dm^{-3}$ = concentration in $mol\,dm^{-3} \times$ RFM	f concentration = number of moles in $25.0\,cm^3 \times 40$ $= 3.44 \times 10^{-3} \times 40 = 0.1376\,mol\,dm^{-3}$
Concentration = 0.1376×40 (RFM of NaOH) = $5.504\,g\,dm^{-3}$	g Concentration in $g\,dm^{-3}$ = concentration in $mol\,dm^{-3} \times$ RFM
	Concentration = 0.1376×40 (RFM of NaOH) = $5.504\,g\,dm^{-3}$

This style of acid–base titration question is common and the only differences may be the ratio in the reaction and whether or not there is a dilution.

Degree of hydration titrations

Titrations can also be used to determine the degree of hydration of a salt but the salt in solution must react with an acid, so hydrated sodium carbonate is the most common example.

Typical degree of hydration titration

- Dissolve hydrated salt (usually hydrated sodium carbonate) in water and make up volume to $250\,cm^3$.
- Pipette $25.0\,cm^3$ of sodium carbonate solution into a conical flask.
- Titrate with 0.1 M hydrochloric acid using methyl orange indicator.
- The average titre can be used to calculate moles of hydrochloric acid:

$Na_2CO_3 + 2HCl \rightarrow 2NaCl + CO_2 + H_2O$

- Calculate the moles of sodium carbonate in $25.0\,cm^3$ of solution using the equation ratio above.
- Calculate the moles of sodium carbonate in $250.0\,cm^3$ of solution.
- Use mass and moles in $250.0\,cm^3$ to calculate RFM of $Na_2CO_3.xH_2O$.
- Subtract 106 for Na_2CO_3 — this will give you RFM of all the water.
- Divide by 18 to calculate moles of water.

Typical degree of hydration calculation question

$2.86\,g$ of hydrated sodium carbonate, $Na_2CO_3.xH_2O$, are dissolved in deionised water, placed in a volumetric flask and the volume made up to $250.0\,cm^3$ using deionised water. A $25.0\,cm^3$ sample of this solution was pipetted into a conical flask and titrated against $0.1\,mol\,dm^{-3}$ hydrochloric acid using methyl orange indicator. The average titre was found to be $20.0\,cm^3$.

Standard types of part questions asked

a Calculate the number of moles of hydrochloric acid added from the burette.

b Write a balanced symbol equation for the reaction between sodium carbonate and hydrochloric acid.

c Calculate the number of moles of sodium carbonate present in $25.0\,cm^3$ of the solution.

d Calculate the number of moles of sodium carbonate present in $250.0\,cm^3$ of the solution.

From this point the calculation can go one of two ways:

RFM method	Ratio method
e Calculate the RFM of hydrated sodium carbonate using the initial mass and the number of moles.	**e** Calculate the mass of anhydrous sodium carbonate present in the sample.
f Calculate the value of x in $Na_2CO_3.xH_2O$.	**f** Calculate the mass of water present in the hydrated sodium carbonate.
	g Calculate the number of moles of water present in the hydrated salt.
	h Calculate the value of x in $Na_2CO_3.xH_2O$.

Answers

a $moles = \dfrac{\text{solution volume} \times \text{concentration}}{1000} = \dfrac{20.0 \times 0.1}{1000} = 0.002\ mol$

b $Na_2CO_3 + 2HCl \rightarrow 2NaCl + H_2O + CO_2$

c 1:2 ratio of Na_2CO_3:HCl so moles of Na_2CO_3 in $25.0\,cm^3 = 0.001\ mol$

d moles of Na_2CO_3 present in $250.0\,cm^3 = 0.001 \times 10 = 0.01\ mol$

Knowledge check 33

$2.67\,g$ of solid $Na_2CO_3.xH_2O$ were dissolved in water and the volume made up to $250\,cm^3$. $25.0\,cm^3$ of this solution reacted with $30.0\,cm^3$ of $0.100\,mol\,dm^{-3}$ hydrochloric acid. Calculate x.

RFM method	Ratio method
e RFM = mass/moles = $\frac{2.86}{0.01}$ = 286	e mass = moles × RFM mass = 0.01 × 106 = 1.06 g
f Na_2CO_3 RFM = 106 RFM of H_2O in $Na_2CO_3.xH_2O$ = 286 – 106 = 180 $x = \frac{\text{RFM of water in hydrated salt}}{\text{RFM of water}}$ $x = \frac{180}{18} = 10$	f mass of H_2O = 2.86 – 1.06 = 1.8 g g moles of H_2O = $\frac{1.8}{18}$ = 0.1 mol h ratio $Na_2CO_3:H_2O$ = 0.01:0.1 = 1:10 so x = 10

Summary

- A titration is carried out using a pipette and a burette and two solutions are mixed to determine the exact volume of one solution required to react with an exact volume of the other solution.
- Questions on acid–base titrations are common where one solution is a base (an alkali) and the other is an acid.
- An indicator is used to determine the exact point when the acid has neutralised the base, or vice versa.
- Two common indicators are phenolphthalein and methyl orange.
- Percentage uncertainty may be calculated for specific apparatus or a calculated value from an experimental procedure.

- The preparation (including rinsings) and accurate use of the apparatus are important in obtaining reliable results.
- The volume of solution added from the burette is called the titre.
- Usually a rough titration is carried out and then subsequent titrations; concordant results (ignoring outliers) are used to calculate the average titre.
- Calculations are carried out based on the average titre to determine concentration, number of moles, mass, RFM, degree of hydration and even to identify unknown elements.

Qualitative tests

Qualitative analysis tests allow for identification of water, gases or ions in an ionic compound.

Tests for gases

Table 18

Gas	Test	Positive result
CO_2	Pass through limewater (calcium hydroxide solution)	Limewater changes from colourless to milky. If CO_2 is in excess, milky white ppt is soluble and forms a colourless solution.
H_2	Put a burning splint into a sample of the gas	Burns with a pop
O_2	Put a glowing splint into a sample of the gas	Splint relights
Cl_2	Damp litmus or universal indicator paper	Bleaches
NH_3	1 Damp universal indicator paper 2 Glass rod dipped in concentrated hydrochloric acid	1 Turns blue 2 Dense white fumes
HCl	1 Damp universal indicator paper 2 Glass rod dipped in concentrated ammonia solution	1 Turns red 2 Dense white fumes

Inorganic identification tests

Exam tip

You must be able to give full practical details of how to carry out the tests for ions. Make sure you include the word 'solution' as this is often left out. 'To a solution of the substance, add silver nitrate solution' — leaving out the term 'solution' here will cost you marks. Remember that a negative test is often as useful as a positive test.

Knowledge check 34

Name the white precipitate formed when silver nitrate solution is added to a solution of sodium chloride.

Table 19

Test and testing for	How to carry out the test	Typical observations	Deductions from observations *Detail of reactions*
1 Appearance	Observe colour and state; state whether crystalline or powder	White crystalline solid	Does not contain transition metal ions/not a transition metal compound Compound of group I, II or ammonium compound
		Blue crystalline solid	Possibly contains copper(II) ions/Cu^{2+} Copper(II) compound
2 Solubility in water	Add a few spatula measures of solid to a test tube of deionised water and shake to mix	Dissolves to form a colourless solution	Soluble in water; group I, II, aluminium, zinc or ammonium compound
		Dissolves to form a solution **State the colour of the solution formed**	Soluble in water If solution is blue suggests copper(II) ions/Cu^{2+} ions present
		Does not dissolve in water	Insoluble in water
3 Flame test Testing for the presence of certain metal cations in the compound	Dip a nichrome wire loop in concentrated hydrochloric acid; touch sample with the wire, then hold it in a blue Bunsen flame	Crimson flame	Lithium ion/Li^+ present
		Yellow/orange flame	Sodium ion/Na^+ present
		Lilac (pink through cobalt blue glass)	Potassium ion/K^+ present
		Brick red/red flame	Calcium ion/Ca^{2+} present
		Green/apple green flame	Barium ion/Ba^{2+} present
		Blue-green/green-blue flame	Copper(II) ion/Cu^{2+} present
5 Silver nitrate solution Testing for halide ions	Dissolve a spatula measure of the solid sample in dilute nitric acid and add a few cm^3 of silver nitrate solution Add dilute/concentrated ammonia solution	White ppt which is soluble in both dilute ammonia solution and concentrated ammonia solution to form a colourless solution	Chloride ion/Cl^- present *White ppt is AgCl*

→

Test and testing for	How to carry out the test	Typical observations	Deductions from observations *Detail of reactions*
		Cream ppt which is not soluble in dilute ammonia solution but is soluble in concentrated ammonia to form a colourless solution	Bromide ion/Br^- present *Cream ppt is AgBr*
		Yellow ppt which is not soluble in either dilute ammonia solution or concentrated ammonia solution	Iodide ions/I^- present *Yellow ppt is AgI*
		No ppt formed	**No** halide ions present
6 Barium chloride solution Testing for sulfate ions	Dissolve a spatula measure of the solid sample in dilute nitric acid (hydrochloric acid also works here) and add a few cm^3 of barium chloride solution	White ppt	Sulfate ions/SO_4^{2-} present *White ppt is BaSO$_4$*
		No ppt formed	**No** sulfate ions/SO_4^{2-} present
7 Dilute acid Testing for carbonate ions The negative part of this test can be the first part of the silver nitrate solution (test 5) or the barium chloride solution (test 6)	Place a few cm^3 of dilute acid in a test tube and add a spatula measure of the solid sample	Effervescence; solid disappears Gas evolved can be passed through limewater in another test tube using a delivery tube — limewater changes from colourless to milky	Carbonate ions/CO_3^{2-} present *Carbon dioxide released from reaction of a carbonate ion with acid*
		No effervescence/no gas produced	**No** carbonate ions/CO_3^{2-} present
12 Sodium hydroxide solution Testing for the ammonium ion/ NH_4^+	Place a few cm^3 of sodium hydroxide solution in a test tube and add a spatula measure of the solid sample; warm gently Test any gas evolved using damp red litmus paper or damp universal indicator paper or test any gas evolved using a glass rod dipped in concentrated hydrochloric acid	Pungent gas evolved Damp indicator paper changes to blue White smoke with glass rod dipped in concentrated hydrochloric acid	Ammonium ion/NH_4^+ present *Ammonia released from action of alkali (NaOH) on an ammonium compound* Ammonia gas is alkaline *White smoke is ammonium chloride from reaction of $NH_3(g)$ with $HCl(g)$ from the concentrated HCl*
13 Starch solution Add starch solution	Add a few cm^3 of starch solution to a sample	Blue-black colour observed	Iodine present
		No blue-black colour observed	No iodine present

Questions & Answers

The AS1 examination is 90 minutes in length and consists of ten multiple-choice questions (each worth 1 mark) and several structured questions, which vary in length. The structured questions make up the remaining 80 marks giving 90 marks in total for the paper. For each multiple choice question there is one correct answer and at least one clear distractor. The mark allocations for the structured questions vary but a general rule is each error loses you a mark.

About this section

This section contains a mix of multiple-choice and structured questions similar to those you can expect to find in the A-level papers.

Each question is followed by brief guidance on how to approach the question and where you could make errors (shown by the ⓔ). Answers to some questions are then followed by comments. These are preceded by the ⓔ. Try the questions first to see how you get on and then check the answers and comments.

General tips

- Be accurate with your learning at this level — examiners will penalise incorrect wording.
- Always follow calculations through to the end, even if you feel you have made a mistake — there are marks for the correct method even if the final answer is incorrect.
- Always attempt to answer a multiple choice question even if it is a guess (you have a 25% chance of getting it right).

The uniform mark you receive for AS1 will be out of 96. Both AS1 and AS2 are awarded out of 96 and the AS3 (practical examination examining practical and planning from AS1 and AS2) is awarded out of 48 uniform marks giving a possible total of 240 for AS chemistry. AS chemistry makes up 40% of the overall A-level.

■ Formulae, equations and amounts of substance

Question 1

Which of the following contains the greatest number of atoms?

A 32 g of sulfur molecules, S_8

B 24 g of magnesium atoms, Mg

C 4 g of hydrogen molecules, H_2

D 31 g of phosphorus molecules, P_4

Questions & Answers

ℓ In this question you can easily calculate the number of particles. Calculate the number of moles using the RFM of the formula given and then multiply by N_A to work out the number of atoms for B and the number of molecules for A, C and D. The number of atoms for A, C and D can be determined by multiplying the number of molecules by the number of atoms in each molecule.

> ## Student answer
>
> **A** $\dfrac{\text{mass (g)}}{\text{RFM}} = \dfrac{32}{256} = 0.125\,\text{mol} \times N_A\ (6.02 \times 10^{23}) = 7.525 \times 10^{22}$ molecules of S_8
>
> Each S_8 contains eight sulfur atoms so number of atoms $= 8 \times 7.525 \times 10^{22}$
> $= 6.02 \times 10^{23}$ atoms of S
>
> **B** $\dfrac{\text{mass (g)}}{\text{RFM}} = \dfrac{24}{24} = 1\,\text{mol} \times N_A\ (6.02 \times 10^{23}) = 6.02 \times 10^{23}$ atoms of Mg
>
> **C** $\dfrac{\text{mass (g)}}{\text{RFM}} = \dfrac{4}{2} = 2\,\text{mol} \times N_A\ (6.02 \times 10^{23}) = 1.204 \times 10^{24}$ molecules of H_2
>
> Each H_2 contains two hydrogen atoms so number of atoms $= 2 \times 1.204 \times 10^{24}$
> $= 2.408 \times 10^{24}$ atoms of H
>
> **D** $\dfrac{\text{mass (g)}}{\text{RFM}} = \dfrac{31}{124} = 0.25\,\text{mol} \times N_A\ (6.02 \times 10^{23}) = 1.505 \times 10^{23}$ molecules of P_4
>
> Each P_4 contains four phosphorus atoms so number of atoms $= 4 \times 1.505 \times 10^{23}$
> 6.02×10^{23} atoms of P
>
> Answer is C ✓

Question 2

Phosphorus, P_4, reacts with bromine to form the colourless liquid phosphorus tribromide, PBr_3.

(a) Write an equation for the reaction of phosphorus, P_4, with bromine to form phosphorus tribromide.

(1 mark)

ℓ When you are given P_4 in the question you will be expected to use it in the equation. The most common mistake would be to use P instead of P_4. Since the question is only worth 1 mark any mistake will lose the mark.

> ## Student answer
>
> **(a)** $P_4 + 6Br_2 \rightarrow 4PBr_3$ ✓

(b) Using the following headings, calculate the mass of phosphorus tribromide formed if 3.1 g of phosphorus is reacted with $10\,\text{cm}^3$ of bromine (density of liquid bromine is $3.1\,\text{g\,cm}^{-3}$).

- mass of bromine, Br_2, in grams
- moles of bromine, Br_2

(1 mark)

(1 mark)

- moles of phosphorus, P_4, in 3.1 g   (1 mark)
- limiting reactant   (1 mark)
- moles of phosphorus tribromide formed   (1 mark)
- mass of phosphorus tribromide formed   (1 mark)

e The headings make this question more approachable because you know how each mark is awarded. Work through it as far as you can. Each step you get right will get you a mark, plus marks in calculations carried through so if you think you have made a mistake, going on with the rest of the question using the correct process but with the wrong numbers will still be credited.

Always scan through the whole question to see where it is leading. In this case 'limiting reactant' and 'density of liquid bromine is 3.1 g cm^{-3}' give you clues as to what type of calculation is expected. If you get a bit lost, try to finish the question with any values you have or simply pick a reactant for limiting and finish the question using this one.

Student answer

(b) Mass of bromine, Br_2 = volume × density = 10 × 3.1 = 31 g ✓

Moles of bromine, Br_2 = 31/160 = 0.194 mol ✓

Moles of P_4 = 3.1/124 = 0.025 mol ✓

Limiting reactant: based on equation $P_4 + 6Br_2 \rightarrow 4PBr_3$

0.025 mol of P_4 react with 0.015 mol of Br_2 *but* 0.194 mol of Br_2 are present so Br_2 is in excess.

P_4 is the limiting reactant. ✓

Based on moles of P_4 (limiting reactant), moles of PBr_3 formed = 0.025 × 4 = 0.1 mol ✓

Mass of PBr_3 formed = mass × RFM = 0.1 × 271 = 27.1 g ✓

Question 3

4.75 g of a hydrated sample of copper(II) sulfate, $CuSO_4.xH_2O$, were heated to constant mass. The mass reduced by 1.71 g. Which of the following is the value of x?

A 3   C 5

B 4   D 6

e The masses that need to be determined in this style of calculation are the mass of the anhydrous copper(II) sulfate and the mass of water lost. Read the question carefully. The decrease in mass on heating to constant mass is the mass of water lost (= 1.71 g). The mass of the anhydrous solid, $CuSO_4$, is the mass remaining after heating to constant mass (= 4.75 – 1.71 = 3.04 g). The moles of anhydrous solid and water can be determined and the simplest ratio worked out.

Student answer

Moles of anhydrous solid, $CuSO_4$ = 3.04/160 = 0.019 mol

Moles of water = 1.71/18 = 0.095 mol

Simplest ratio of $CuSO_4$:H_2O = 1:5 so $x = 5$

Answer is C ✓

Atomic structure

Question 1

Which of the following is the electronic configuration of a titanium ion, Ti^{2+}?

A $1s^2\ 2s^2\ 2p^6\ 3s^2\ 3p^6\ 4s^2$

B $1s^2\ 2s^2\ 2p^6\ 3s^2\ 3p^6\ 3d^2\ 4s^2$

C $1s^2\ 2s^2\ 2p^6\ 3s^2\ 3p^6\ 3d^4$

D $1s^2\ 2s^2\ 2p^6\ 3s^2\ 3p^6\ 3d^2$

ⓔ The atomic number of titanium is 22, so a titanium atom has 22 protons and 22 electrons. The Ti^{2+} ion has lost two electrons and remember that transition metal atoms lose 4*s* electrons first. The electronic configuration of a titanium ion Ti^{2+} is $1s^2\ 2s^2\ 2p^6\ 3s^2\ 3p^6\ 3d^2\ 4s^2$ — the $4s^2$ electrons were lost.

Student answer

Answer is D ✓

Question 2

(a) Write an equation for the first ionisation energy of potassium, including state symbols. (2 marks)

Student answer

(a) $K(g) \rightarrow K^+(g) + e^-$ ✓✓

ⓔ 1 mark is awarded for the correct equation and 1 mark for the correct state symbols. To answer this, remember the definition of first ionisation energy — the key points are gaseous atoms losing 1 mole of electrons and forming gaseous ions with a single positive charge. The most common error in this answer would be incorrect state symbols for the potassium atom and the ion. Some answers might attempt to include a state symbol for the electron. Electrons are lost so should appear on the right-hand side of the equation.

(b) Explain why the first ionisation energy of rubidium is smaller than the first ionisation energy of potassium.

> **(b)** Outer electron for rubidium is further from the nucleus/atomic radius increases from potassium to rubidium. ✓
>
> Rubidium's outer electron is shielded by more inner shells of electrons. ✓

ℯ The main point in any question on differences in ionisation energies is to identify which of the four factors that affect ionisation energies apply to this answer. They are atomic radius, (effective) nuclear charge, shielding by inner electrons and stability of filled and half-filled subshells. Both potassium and rubidium have outer shell s^1 electronic configurations, so there is no difference in stability of filled or half-filled subshells. The (effective) nuclear charge is the same as both atoms have equal numbers of protons and electrons, so net attraction is the same. For rubidium it is the increasing atomic radius (distance from the nucleus) and more shielding by inner electrons that cause the lower first ionisation energy.

Question 3

The mass spectrum of zirconium (atomic number 40) indicates five different isotopes, which have the following relative abundances.

Relative isotopic mass	Relative abundance
90	51.5
91	11.2
92	17.1
94	17.4
96	2.8

Calculate the relative atomic mass of zirconium to two decimal places. (3 marks)

ℯ When calculating relative atomic mass from relative isotopic masses data or from a mass spectrum directly, you should multiply the mass by the relative abundance for each isotope, then add them all up. Finally divide by the total of all the relative abundances. If the question asks for a specific number of decimal places, stick to this or you will lose a mark.

> **Student answer**
> Relative atomic mass = $\dfrac{(90 \times 51.5) + (91 \times 11.2) + (92 \times 17.1) + (94 \times 17.4) + (96 \times 2.8)}{51.5 + 11.2 + 17.1 + 17.4 + 2.8}$ ✓
> = 9131.8/100 = 91.318 ✓ = 91.32 ✓ to 2 decimal places

■ Bonding

Question 1

Chlorine reacts with magnesium to form magnesium chloride. Magnesium chloride is soluble in water and magnesium chloride solution conducts electricity.

(a) Using a dot and cross diagram, explain how atoms of magnesium react with atoms of chlorine to form magnesium chloride. Show outer electrons only.   (4 marks)

Student answer

(a)

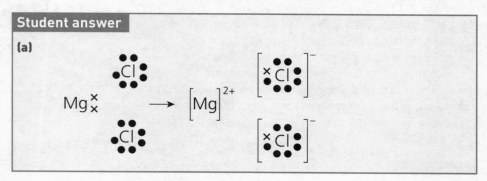

e Magnesium chloride is an ionic compound. Only show outer electrons in the answer. Magnesium atoms have two outer electrons and chlorine atoms have seven outer electrons.

One magnesium atom requires two chlorine atoms. Draw all the atoms showing a cross (x) for electrons on one type of atom and a dot (●) for the electrons on the other type of atom. Each of the two outer electrons of the magnesium atom transfers to each chlorine atom (still draw as an × to show where they came from). The magnesium ion now has no electrons in this outer shell and each chloride ion has eight electrons in its outer shell. Show the charges on the ions.

- Correct electronic configuration of Mg atom ✓
- Correct electronic configuration of Cl atom ✓
- Correct number of each atom/ion, i.e. 1:2 ratio ✓
- Correct electron transfer and charges on the ions ✓

(b) Explain why a solution of magnesium chloride conducts electricity.   (2 marks)

(b) Magnesium chloride is an ionic compound and when dissolved in water the ions ✓ are free to move and carry charge. ✓

e The most common error here is to confuse ions and electrons. Metals and graphite conduct electricity due to the delocalised electrons, which are free to move and carry charge, but molten ionic compounds, ionic compounds dissolved in water and acids all conduct electricity due to the free ions that can move and carry charge. You will lose a mark if you confuse the charged particles that can move.

Question 2

Beryllium chloride is a covalently bonded molecule as shown in the dot-and-cross diagram.

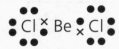

Beryllium chloride

(a) Explain why beryllium chloride does not obey the octet rule. (2 marks)

🄮 The octet rule states that when forming a compound an atom will lose, gain or share electrons to achieve eight electrons in its outer shell. You must state the octet rule and then explain which atom(s) obey or do not obey it and how they do not obey it. In this example it is the beryllium atom that does not obey the octet rule but you must explain clearly how (Be has only four electrons in its outer shell).

> **Student answer**
>
> **(a)** When forming a compound an atom will lose, gain or share electrons to achieve eight electrons in its outer shell ✓
>
> The Be atom does not obey the octet rule because it has only four electrons in its outer shell. ✓

(b) Explain the term electronegativity. (1 mark)

🄮 This is a common definition style question and it is important that you include all aspects of the definition to gain the marks. Do not change it, and make sure you use the proper scientific terms as often one wrong word can cost you a mark.

> **(b)** The extent to which an atom attracts the bonding electrons in a covalent bond. ✓

(c) Label the diagram below to show the polarity of the Be–Cl bond. (1 mark)

Be–Cl

🄮 This question is essentially asking about the trends in electronegativity of elements. The more electronegative atom will be $\delta-$ and the other $\delta+$. Atoms that are further to the right in the periodic table are more electronegative and those further up the periodic table are more electronegative. Cl is much further to the right and so is more electronegative, so Cl gets the $\delta-$ and Be the $\delta+$. Make sure you write them above the atoms to indicate the polarity.

> **(c)** $\delta+ \quad \delta-$
> Be–Cl ✓

(d) Explain why beryllium chloride is a non-polar molecule even though it contains polar bonds. (2 marks)

🄮 This is a common question as many molecules are non-polar even though they contain polar bonds. The molecules like this must contain equally polar bonds, which are arranged symmetrically so the polarities (dipoles) of these bonds cancel each other out. Common examples are CO_2, CCl_4, BF_3 and SF_6. Similar molecules like CS_2, CF_4 and BCl_3 would also be non-polar for the same reasons.

> **(d)** Beryllium chloride contains equally polar bonds which are arranged symmetrically ✓ and so the polarities of the bonds (dipoles) cancel each other out ✓

Question 3

Which one of the following contains a coordinate bond?

A $CaCl_2$ **C** HCl

B CCl_4 **D** NH_4Cl

ⓔ $CaCl_2$ is ionic, CCl_4 is a molecular covalent substance with no coordinate bonds, HCl is a covalent molecule with a single covalent bond between the atoms, and NH_4Cl is an ionic substance that contains the ammonium ion, which has a coordinate bond in its structure.

> Answer is D ✓

▮Intermolecular forces

Question 1

Which of the following liquids is polar?

A CCl_4 **C** C_2H_5OH

B CS_2 **D** C_6H_{14}

ⓔ This type of question can be asked in many different ways and the molecules chosen can vary. Again, use your knowledge of shape and polarity to decide which are the non-polar molecules. CS_2 will be non-polar based on CO_2 being non-polar. All hydrocarbons are non-polar. CCl_4 and CS_2 both contain polar bonds but the bonds are arranged symmetrically so the polarities cancel out, making the molecules non-polar overall. The question could ask 'a stream of which liquid would be deflected by a charged rod'. A stream of polar liquid would be deflected by a charged rod.

> **Student answer**
> Answer is C ✓

Question 2

Iodine is a molecular covalent crystalline solid that has van der Waals forces of attraction. Van der Waals forces are a type of intermolecular bonding.

(a) State one other type of intermolecular bonding. (1 mark)

(b) Explain how van der Waals forces of attraction arise between molecules. (2 marks)

e There are three types of intermolecular forces between simple covalent molecules: van der Waals forces of attraction, permanent dipole attractions and hydrogen bonds. You need to be able to recognise which forces of attraction occur between specific molecules. Remember that non-polar molecules (such as I_2) only have van der Waals forces of attraction between their molecules. Polar molecules will have permanent dipole attractions and if a polar molecule has a hydrogen atom bonded to an N, O or F atom, this H atom can form a hydrogen bond with a very electronegative atom with a lone pair of electrons. Make sure you can explain how the intermolecular forces are formed.

Student answer

(a) Hydrogen bonds/permanent dipole attractions ✓

(b) Temporary dipoles ✓ caused by random movement of electrons ✓

Question 3

Explain why water has a higher boiling point than hydrogen sulfide. (3 marks)

e Any question about physical state or melting/boiling points of simple covalent substances is linked to intermolecular forces. It is vital that you can correctly identify a substance as simple covalent and then identify the intermolecular forces that are needed to explain the properties. Water forms stronger hydrogen bonds between its molecules as well as van der Waals forces of attraction whereas hydrogen sulfide, H_2S, has weaker van der Waals forces of attraction and permanent dipole attractions. The answer must give the type of intermolecular bonding and say which is stronger. Give a full answer, bringing in the idea of the energy required to break the bonds to ensure you gain all the marks.

Student answer

Hydrogen bonds between water molecules ✓ are stronger than van der Waals forces of attraction/permanent dipole attraction in H_2S. ✓

More energy is needed to break the stronger hydrogen bonds. ✓

■ Structure

Question 1

Which of the following does not have a molecular covalent crystalline structure?

A   diamond   C   iodine

B   ice   D   rhombic sulfur

ⓔ This is a 'negative' question. Highlight the *not* in the question to remind you. Describe the structure of each and decide which one does not exist as molecular covalent crystals. All are crystalline solids but diamond is giant covalent; the rest are molecular covalent.

Student answer

Answer is A ✓

Question 2

One form of boron nitride has a structure similar to that of graphite.

(a) What is the chemical formula for boron nitride?   (1 mark)

(b) Describe the structure and bonding of boron nitride.   (2 marks)

(c) Suggest why boron nitride does not conduct electricity.   (1 mark)

ⓔ The boron and nitrogen atoms in boron nitride can both form three covalent bonds, so it can form a hexagonal layered arrangement similar to graphite with alternating boron and nitrogen atoms. It has the same lubricant properties of graphite but it does not conduct electricity because it does not have any delocalised electrons to move and carry charge. The three electrons in the outer shell of the boron atom are used in bonding and the lone pair of electrons cannot be delocalised.

Student answer

(a) BN ✓

(b) Bonding is covalent ✓ Structure is giant covalent/macromolecular ✓

(c) No delocalised electrons that can move and carry charge ✓

Question 3

The table below details some molecular covalent substances and some giant covalent substances.

Substance	Melting point/°C	Boiling point/°C	Electrical conductivity
A	3550	4827	Does not conduct
B	44	281	Does not conduct
C	3677	4050	Conducts
D	1610	2230	Does not conduct
E	105	183	Does not conduct

Using the information in the table above, answer the following questions.

(a) Which substance is graphite? Explain your answer.   (2 marks)

(b) Which two of the substances are molecular covalent crystals?   (2 marks)

(c) Which of the substances is diamond?   (1 mark)

ⓔ This question uses your knowledge of the properties of giant covalent and molecular covalent substances. Giant covalent substances have very high melting points and boiling points and graphite conducts electricity whereas diamond does not. Molecular covalent or simple covalent substances have low melting points and do not conduct electricity. The last question is slightly more tricky. You would be expected to choose between A and D. The melting point of D is not high enough to be diamond since diamond and graphite have similar melting points over 3500°C.

Student answer

(a) C ✓ high melting point and conducts electricity ✓

(b) B ✓ and E ✓

(c) A ✓

Shapes of molecules and ions

Question 1

Which of the following gives the correct shapes of the molecules ammonia, water and carbon dioxide?

	Ammonia shape	Water shape	Carbon dioxide shape
A	Tetrahedral	Linear	Bent
B	Pyramidal	Linear	Linear
C	Trigonal planar	Bent	Linear
D	Pyramidal	Bent	Linear

ⓔ It is important to remember the shapes of various molecules. Water is often wrongly identified as linear and ammonia as trigonal planar. Carbon dioxide is linear due to the equal repulsion of the bonding electrons in the double bonds. Remember that molecules with two bonds can be linear or bent; molecules with three bonds can be trigonal planar or pyramidal.

Student answer

Answer is D ✓

Question 2

The diagram below is a dot-and-cross diagram for phosphine, PH_3.

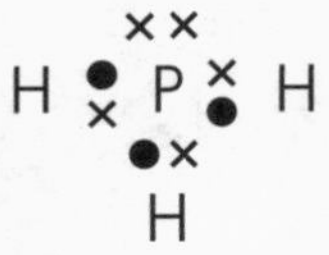

$$H \overset{\times\times}{\underset{\times}{\bullet}} P \overset{\times}{\underset{\bullet\times}{\bullet}} H$$
$$H$$

State and explain the shape of a molecule of phosphine.   (4 marks)

ⓔ It is common to base such a question on a familiar compound but change the central atom to that of another element in the same group. PH_3 is similar to NH_3. The explanation of its shape is the same as the explanation for NH_3. This can also be done with H_2S (similar to H_2O), SiH_4 (similar to CH_4) etc.

> **Student answer**
> - Shape of PH_3 is pyramidal ✓
> - Electron pairs repel each other ✓
> - Three bonding pairs of electrons and one lone pair of electrons and lone pair of electrons has greater repulsion✓
> - Molecule takes up shape to minimise repulsions ✓

Question 3

Which of the following is the shape of the ClF_4^- ion?

A octahedral

B square planar

C tetrahedral

D trigonal bipyramidal

ⓔ Chlorine has seven electrons in its outer shell. The negative charge gives it eight. Four of these electrons are used to form the covalent bonds with the fluorine atoms. This leaves four electrons, which exist as two lone pairs. There are therefore six pairs of electrons — four bonding pairs and two lone pairs. The shape is similar to XeF_4, which is square planar with the lone pairs taking the up and down positions. Octahedral would require six bonding pairs of electrons; tetrahedral would require four bonding pairs of electrons only; trigonal bipyramidal would require five bonding pairs of electrons only.

> **Student answer**
> Answer is B ✓

■ Redox

Question 1

Explain, in terms of oxidation numbers, why the following reaction is described as a redox reaction.   (3 marks)

$$6FeSO_4 + 3Cl_2 \rightarrow 2Fe_2(SO_4)_3 + 2FeCl_3$$

> **Student answer**
> - Iron/Fe is oxidised from +2 (in $FeSO_4$) to +3 (in $Fe_2(SO_4)_{3 \text{ and } FeCl3}$). ✓
> - Chlorine is reduced from 0 (in Cl_2) to –1 (in $FeCl_3$). ✓
> - Redox is oxidation and reduction occurring in the same reaction. ✓

(e) Most answers will achieve the final mark but many will confuse the calculation of the oxidation states. Remember that the sulfate ion is SO_4^{2-} so the iron (Fe) in $FeSO_4$ has an oxidation state of +2 and the iron (Fe) in both $Fe_2(SO_4)_3$ and $FeCl_3$ has an oxidation state of +3. All compounds have an overall oxidation state of 0. All elements, even diatomic ones like chlorine, Cl_2, have an oxidation state of 0. The balancing numbers in the equation do not affect the oxidation state. The sulfate ion, SO_4^{2-}, does not change so the oxidation state of the sulfur or oxygen also does not change.

Question 2

For the following half-equation, *a*, *b* and *c* are the number of moles of water, hydrogen ions and electrons required respectively.

$$SO_2 + aH_2O \rightarrow SO_4^{2-} + bH^+ + ce^-$$

Which one of the following is correct?

	a	b	c
A	1	2	2
B	2	2	2
C	2	2	4
D	2	4	2

(e) To answer this question, calculate the oxidation state of sulfur in SO_2 (+4) and in SO_4^{2-} (+6). The equation is an oxidation because the electrons are being lost (on the right-hand side of the equation), so there must be an increase in oxidation state. Two moles of water are required on the left to supply the extra oxygen atoms and $4H^+$ must be a product. Practise balancing these equations but watch out for the dichromate(VI) one where the ratio of dichromate to Cr^{3+} is 1:2.

Student answer

Answer is D ✓

Question 3

Using the half-equations shown below:

$$MnO_4^- + 8H^+ + 5e^- \rightarrow Mn^{2+} + 4H_2O$$
$$2Br^- \rightarrow Br_2 + 2e^-$$

write an ionic equation for the oxidation of bromide ions using acidified potassium manganate(VII). (2 marks)

(e) Before tackling this question it is important to remember that an ionic equation does not contain electrons. You must combine an oxidation and reduction half-equation, multiplying each if necessary so that the electrons cancel out on each side. Remember, for more complex ionic equations you may have to cancel down H^+ ions and H_2O as well to simplify the final equation. In this example, multiply the manganate(VII) half-equation by 2 and multiply the bromide half-equation by 5. Then add the equations together by writing down everything on the left of both arrows then putting an arrow before writing down everything on the right of the arrows. Then cancel out anything that appears on both sides (in this case $10e^-$) and you have the finished ionic equation.

Questions & Answers

Student answer

$2MnO_4 + 16H^+ + 10e^- \rightarrow 2Mn^{2+} + 8H_2O$

$10Br^- \rightarrow 5Br_2 + 10e^-$

$2MnO_4^- + 16H^+ + \cancel{10e^-} + 10Br^- \rightarrow 2Mn^{2+} + 8H_2O + 5Br_2 + \cancel{10e^-}$

Ionic equation:

$2MnO_4^- + 16H^+ + 10Br^- \rightarrow 2Mn^{2+} + 8H_2O + 5Br_2$ ✓✓

■Halogens

Question 1

Sodium iodide reacts with concentrated sulfuric acid.

(a) Name all the products of the reaction. (3 marks)

(b) State what would be observed during the reaction. (3 marks)

(c) Write two balanced symbol equations for the reactions that are occurring. (2 marks)

e Remember that the equations for the reactions of the halides with concentrated sulfuric acid are very similar, with the first equation being common to the fluoride, chloride, bromide and iodide; the second is common to the bromide and iodide. The last two equations must be recalled for the iodide but knowing the equations allows you to name all the products as well as stating observations for the reactions. So it is best to answer part (c) first and use the equations to answer parts (a) and (b). If the equations are not asked for, write them down to remind yourself. You will not lose marks for incorrect equations if they are not asked for.

Student answer

(c) $NaI + H_2SO_4 \rightarrow HI + NaHSO_4$

$2HI + H_2SO_4 \rightarrow I_2 + SO_2 + 2H_2O$

$6HI + H_2SO_4 \rightarrow 3I_2 + S + 4H_2O$

$8HI + H_2SO_4 \rightarrow 4I_2 + H_2S + 4H_2O$

e Any two score 1 mark each.

(a) hydrogen iodide; sodium hydrogensulfate; iodine; sulfur dioxide; water; sulfur; hydrogen sulfide ✓✓✓

e For all seven products named correctly 3 marks are awarded. Six products named correctly score 2 marks. Five products named correctly score 1 mark. Four or fewer products named correctly score 0 marks.

(b) *Any three from*:

■ misty/steamy fumes

■ pungent gas

■ purple vapour or grey-black solid

■ yellow solid

■ rotten eggs smell

■ heat released ✓✓✓

Question 2

State the colour of each of the following:

(a) NaBr(s)

(c) KI(aq)

(b) $Cl_2(g)$

(d) I_2 (dissolved in hexane)

(1 mark each)

Student answer

(a) White ✓

(b) Yellow-green ✓ (or green-yellow or green)

(c) Colourless ✓

(d) Purple ✓

e Questions concerning the colours of the halogens and halides are common and you should learn these thoroughly. Remember that group I and II halides are white solids that dissolve in water forming colourless solutions. Most of these questions are recall (AO1) and these are marks you cannot afford to lose. Often they may be asked as observations for reactions, but knowing the colours can help you deduce what should be observed.

Question 3

Chlorine reacts with water and with sodium hydroxide solution. The following compounds of chlorine are formed:

NaCl   NaOCl $NaClO_3$ HOCl   HCl

(a) Name the highest oxidation state compound of chlorine formed in the reaction of chlorine with hot concentrated sodium hydroxide solution. (1 mark)

(b) Write an equation for the reaction of chlorine with cold dilute sodium hydroxide solution. (1 mark)

(c) Name the two products of the reaction of chlorine with water. (2 marks)

(d) Explain why the reaction of chlorine with water is described as disproportionation. (2 marks)

e It is important to be able to identify the oxidation and reduction products of the reactions of chlorine with water and with sodium hydroxide solution. Make sure you can name the products properly, including the oxidation state of chlorine. There is a difference in the reaction of chlorine with cold dilute sodium hydroxide solution and with hot concentrated sodium hydroxide solution. You should know the equations and be able to name the products. These reactions are all disproportionation so make sure you can explain it in terms of oxidation state, and also know the definition of disproportionation.

Student answer

(a) Sodium chlorate(v) ✓

(b) $NaOH + Cl_2 \rightarrow NaOCl + NaCl + H_2O$ ✓

(c) Hydrochloric acid ✓ and hypochlorous acid/chloric(I) acid ✓

(d) Chlorine oxidised from 0 to +1 and reduced from 0 to –1 ✓

Same element is oxidised and reduced in one reaction ✓

■ Acid-base titrations

Question 1

Which one of the following shows the most suitable indicator for the titration?

	Titration	Indicator
A	Ethanoic acid and sodium hydroxide solution	Methyl orange
B	Nitric acid and ammonia solution	Phenolphthalein
C	Sodium hydroxide solution and sulfuric acid	Phenolphthalein
D	Ammonia solution and ethanoic acid	Methyl orange

ⓔ The choice in an indicator in a titration depends on the acid and alkali involved. Strong acids (hydrochloric acid, sulfuric acid and nitric acid) reacting with a strong alkali (sodium hydroxide solution or potassium hydroxide solution) can use either phenolphthalein or methyl orange. A strong acid with a weak alkali (mainly ammonia solution or sodium carbonate solution) must use methyl orange. A weak acid (mostly organic acids like ethanoic acid) with a strong alkali must use phenolphthalein and, finally, for a weak acid with a weak alkali there is no suitable indicator and a pH meter is the most suitable way of monitoring such a titration.

Student answer

Answer is C ✓

Question 2

20.0 cm³ of 0.03 M sulfuric acid is exactly neutralised by:

A   15.0 cm^3 of 0.02 M sodium hydroxide solution

B   15.0 cm^3 of 0.04 M sodium hydroxide solution

C   30.0 cm^3 of 0.02 M sodium hydroxide solution

D   30.0 cm^3 of 0.04 M sodium hydroxide solution

🄮 The most common error in this question would be to assume a 1:1 ratio in the reaction between sodium hydroxide and sulfuric acid. Always write the balanced symbol equation for the titration reaction to check the ratio:

$2NaOH + H_2SO_4 \rightarrow Na_2SO_4 + 2H_2O$

The ratio of NaOH to H_2SO_4 is 2:1. 20.0 cm^3 of 0.03 M sulfuric acid is 0.0006 mol. This reacts with 0.0012 mol of NaOH. Which of the answers A to D gives 0.0012 mol of NaOH?

Student answer

Answer is D ✓

Question 3

0.715 g of a sample of hydrated sodium carbonate, $Na_2CO_3.xH_2O$, were dissolved in deionised water and the volume of the solution was made up to 250 cm^3 in a volumetric flask.

25.0 cm^3 of this sample were titrated against 0.05 mol dm^{-3} hydrochloric acid using methyl orange indicator. The average titre was 12.3 cm^3.

(a) State the colour change observed at the end point. (2 marks)

(b) Calculate the number of moles of hydrochloric acid used in this titration. (1 mark)

(c) Write the equation for the reaction between sodium carbonate and hydrochloric acid. (2 marks)

(d) Calculate the number of moles of sodium carbonate present in 25.0 cm^3 of solution. (1 mark)

(e) Calculate the number of moles of sodium carbonate present in 250 cm^3 of solution. (1 mark)

(f) Calculate the mass of sodium carbonate, Na_2CO_3, present in the sample. (1 mark)

(g) Calculate the mass of water present in the sample of hydrated sodium carbonate. (1 mark)

(h) Calculate the number of moles of water present in the sample. (1 mark)

(i) Calculate the value of x in $Na_2CO_3.xH_2O$. (1 mark)

🄮 This is a standard type of structured titration question. The question leads you through the answer and guides you to determine the moles of the anhydrous salt and the moles of water and then the value of x (degree of hydration) by determining the simplest ratio of anhydrous salt to water.

If is often useful to do a sketch of the titration to make sure you know what is happening at each stage. This can help you work out the colour change of the indicator and also carry out the calculation. For example:

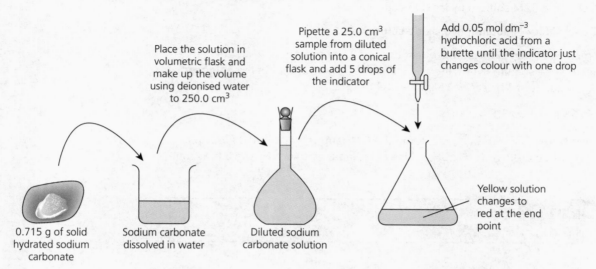

Place the solution in volumetric flask and make up the volume using deionised water to 250.0 cm³

Pipette a 25.0 cm³ sample from diluted solution into a conical flask and add 5 drops of the indicator

Add 0.05 mol dm⁻³ hydrochloric acid from a burette until the indicator just changes colour with one drop

Yellow solution changes to red at the end point

0.715 g of solid hydrated sodium carbonate

Sodium carbonate dissolved in water

Diluted sodium carbonate solution

Remember when determining the colour change at the end point that the acid is added to the alkali. This is clear as the average titre is for hydrochloric acid and the average titre is for the solution added from the burette.

Student answer

(a) yellow ✓ to red ✓

e Even if you cannot work out which solution is being added from the burette, make an educated guess at the colour change for methyl orange as even the wrong way round will gain 1 mark if 2 marks are on offer.

(b) moles of HCl = $\dfrac{\text{solution volume (cm}^3\text{)} \times \text{concentration (mol dm}^{-3}\text{)}}{1000}$

$\quad = \dfrac{12.3 \times 0.05}{1000}$

$\quad = 0.000615 \text{ mol} ✓$

e The first step in most titration questions is to calculate the number of moles of the solute in the solution added from the burette. Make sure you do not confuse the solution volumes — 25.0 cm³ was added to the conical flask and 12.3 cm³ was added from the burette. Check twice that it is the correct solution.

(c) $Na_2CO_3 + 2HCl \rightarrow 2NaCl + CO_2 + H_2O$ ✓✓

ⓔ The most common mistake in this equation is to miss the 1:2 ratio of Na_2CO_3:HCl. Practise writing these equations where carbonates of group I react with strong acids.

(d) 1:2 ratio of Na_2CO_3:HCl so the moles of $Na_2CO_3 = \dfrac{\text{moles of HCl}}{2}$

$= \dfrac{0.000615}{2} = 0.0003075\,\text{mol} \checkmark$

ⓔ The second step in a titration calculation is usually working out the number of moles of solute dissolved in the solution volume (usually $25.0\,\text{cm}^3$) that was pipetted into the conical flask. The ratio of the reaction between the two solutes allows you to do this. 1 mol of Na_2CO_3 reacts with 2 mol of HCl so to move from HCl to Na_2CO_3 you need to divide the number of moles by 2. Think this step out logically each time. Errors are often made here with students multiplying by 2 instead of dividing by 2.

(e) $0.0003075 \times 10 = 0.003075\,\text{mol} \checkmark$

ⓔ The number of moles of this solute (Na_2CO_3) in $25.0\,\text{cm}^3$ of solution is multiplied by 10 to determine the number of moles of the solute in $250\,\text{cm}^3$ of the solution (since a $25.0\,\text{cm}^3$ sample of the solution was taken, so 1/10th of the moles were taken from the volumetric flask). This is the number of moles of Na_2CO_3 in solution and is the same as the number of moles of solid $Na_2CO_3.xH_2O$ added to the solution. It is important to realise the distinction between solid $Na_2CO_3.xH_2O$ which is hydrated and sodium carbonate in solution, which is simple $Na_2CO_3(aq)$. The water of crystallisation is not part of the mass of the solute in solution but the number of moles of each is the same.

(f) $0.003075 \times 106 = 0.326\,\text{g} \checkmark$

ⓔ The mass of Na_2CO_3 dissolved refers to the mass of Na_2CO_3 (not $Na_2CO_3.xH_2O$). You do not know the degree of hydration so you cannot work out an RFM of $Na_2CO_3.xH_2O$. As with all moles–mass calculations, multiply the number of moles by the RFM (in this case 106 for Na_2CO_3).

(g) $0.715 - 0.326 = 0.389\,\text{g} \checkmark$

ⓔ $0.715\,\text{g}$ of solid $Na_2CO_3.xH_2O$ were added but only $0.326\,\text{g}$ were determined to be Na_2CO_3. The rest of the mass must be the xH_2O. Subtraction gives the mass of water in $Na_2CO_3.xH_2O$.

(h) $\dfrac{0.389}{18} = 0.0216\,\text{mol} \checkmark$

ⓔ As with all mass–moles calculations, divide the mass by the RFM. The RFM of water is 18 in all calculations.

> **(i)** ratio of Na_2CO_3:H_2O = 0.003075:0.0216 = 1:7.02 = 1:7 so x = 7

ⓔ Using the moles of Na_2CO_3 determined in part (e) and the moles of water determined in part (h) divided by the smallest number of moles (0.003075) to reduce the ratio to 1:x, the value of x is determined to be 7.02. Rounding errors during the calculation and also during the titration may have resulted in a rough answer, so the answer provided should be the nearest whole number, x = 7. Note that x does not have to be a whole number because hydrated salts lose water of crystallisation gradually over time, so the actual value may be between 7 and 8, but this should be indicated in the question.

Question 4

Describe, giving practical details, how you would accurately prepare a 250 cm³ solution containing 0.715 g of hydrated sodium carbonate.

(4 marks)

> **Student answer**
> - Weigh accurately 0.715 g of hydrated sodium carbonate in a container. ✓
> - Add (a minimum quantity of) deionised water and stir with a glass rod until dissolved. ✓
> - Add the solution to a 250 cm³ volumetric flask using a filter funnel. ✓
> - Rinse the rod and container (and funnel) into a 250 cm³ volumetric flask through a filter funnel, ensuring all rinsings enter the flask. ✓
> - Add deionised water until the bottom of the meniscus is on the line. ✓
> - Mix the solution by stoppering the flask and inverting it several times. ✓
> (*maximum 4 marks*)

ⓔ This type of question is common in AS and A2 practical exams as well as in AS Unit 1 papers. Make sure you include all the rinsings to make up the solution as well as the idea of the bottom of the meniscus being on the line in the volumetric flask. Practical detail is key here.

■ Qualitative tests

Question 1

A solid compound was dissolved in deionised water. Dilute nitric acid was added followed by silver nitrate solution. A white precipitate was observed which dissolved when ammonia solution was added.

(a) Name the ion that has been identified during this test. (1 mark)

(b) Write an ionic equation for the formation of the white precipitate. (1 mark)

(c) Suggest why dilute nitric acid is added to the solution before silver nitrate solution. (2 marks)

e Silver nitrate solution is the key to this question. It is used to test for halide (chloride, bromide and iodide) ions. The dilute nitric acid is added to remove any carbonate ions that would give a white precipitate of silver carbonate, Ag_2CO_3. This would be a false positive test for chloride ions. Learn the observations carefully in terms of the colour of the precipitates and whether they are soluble in ammonia solution.

Student answer

(a) Chloride ✓

(b) $Ag^+ + Cl^- \rightarrow AgCl$ ✓

(c) Nitric acid reacts with/removes carbonate ions. ✓ Carbonate ions would give a white precipitate/false test for chloride ions. ✓

Question 2

A mixture of two compounds (Y and Z) was dissolved in water and tested. The results are shown in the table.

Flame test on solid mixture	Lilac colour
Addition of dilute nitric acid to solid mixture	Effervescence
Addition of silver nitrate solution	Yellow precipitate

Y and Z could be:

A potassium carbonate and potassium sulfate

B potassium chloride and potassium carbonate

C potassium iodide and potassium carbonate

D sodium iodide and sodium carbonate

Student answer

Answer is C ✓

---

e The use of a mixture of compounds is common in the AS practical examination. You would be expected to identify both compounds in a mixture from a series of tests but it can also be asked in an AS Unit 1 paper. The lilac colour indicates the presence of K^+ ions. The effervescence indicates carbonate ions being present. The yellow precipitate with silver nitrate solution tells us that there are iodide ions present. If you were describing this in a written prose style question, be careful to use the terms 'solution' and 'ions'. 'Potassium is present' implies potassium metal is present instead of potassium ions. 'Adding silver nitrate' implies adding solid silver nitrate but when observing precipitates it is two solutions that are mixed.

Question 3

Describe how you would show the presence of the sodium ion and the bromide ion in a sample of solid sodium iodide.

(8 marks)

ⓔ This type of question involves a large amount of practical detail. You would be expected to explain how to make up a solution of the solid, how to carry out a flame test, and the solutions required and observations for a positive test for iodide ions. The colour of the flame test for sodium ions is either yellow or orange but both should not be given. It would be expected that the word solution is used with every solution except dilute nitric acid. The addition of the dilute nitric acid removes carbonate ions, which may give a false positive for the presence of chloride ions since silver carbonate is a white solid. The yellow precipitate is silver iodide. It is important to note that it does not dissolve in either dilute or concentrated ammonia solution.

Student answer

- Dip a piece of nichrome wire in concentrated hydrochloric acid. ✓
- Place in the solid sample and into a blue Bunsen burner flame. ✓
- A yellow/orange flame indicates the presence of sodium ions. ✓
- Dissolve the solid in deionised water/make a solution of the solid. ✓
- Add dilute nitric acid ✓ followed by silver nitrate solution. ✓
- A yellow precipitate forms ✓ which does not dissolve in dilute or concentrated ammonia solution. ✓

Knowledge check answers

1 $(NH_4)_2Cr_2O_7$
2 $Mg_3N_2 + 6H_2O \rightarrow 3Mg(OH)_2 + 2NH_3$
3 0.0151 mol
4 1.51×10^{22}
5 0.902 g
6 32.9 g
7 $x = 7$
8 Proton: relative mass = 1; relative charge = +1
   Electron: relative mass = 1/1840; relative charge = –1
   Neutron: relative mass = 1; relative charge = 0
9 The molecular ion is the ion formed by the removal of an electron from a molecule.
10 Outer shell electrons are in the **s** subshell.
11 Atomic radius; nuclear charge; shielding by inner electrons; stability of filled and half-filled subshells.
12 Ionic
13 Attraction between positive ions in lattice and sea of delocalised electrons.
14 Electrostatic attraction between a shared pair of electrons and the nuclei of the bonded atoms.
15 Carbon in methane obeys the octet rule as eight electrons in its outer shell.
16 Chlorine is more electronegative as it is higher in group VII.
17 Hydrogen bonds/permanent dipole–dipole attractions/van der Waals forces.
18 Fixed hydrogen bonds lead to a more open structure in ice so a lower density.
19 Delocalised electrons between the layers can move and carry charge.
20 Trigonal planar 120°
21 Octahedral 90°
22 Tetrahedral and octahedral
23 +5
24 $SO_3^{2-} + 6H^+ + 4e^- \rightarrow S + 3H_2O$
25 Same element is oxidised and reduced in the same reaction.
26 Sodium chlorate(I)
27 Chlorine
28 Iodide
29 $KBr + H_2SO_4 \rightarrow KHSO_4 + HBr$
   $2HBr + H_2SO_4 \rightarrow Br_2 + SO_2 + 2H_2O$
30 40
31 Hydrochloric acid/sulfuric acid/nitric acid
32 Pink to colourless
33 $x = 4$
34 Silver chloride

Index

Note: page numbers in bold indicate defined terms.

Index